·RADIO TELEPHONY·

BY

ALFRED N. GOLDSMITH, Ph. D.

Fellow of the Institute of Radio Engineers
Member of the American Institute of Electrical Engineers

Director of the Radio
Telegraphic and Telephonic Laboratory
and Professor at
The College of the City of New York

THE WIRELESS PRESS, Inc.
25 Elm Street, New York

CONTENTS

(Classified Topically and According to Chapters)

III

PREFACE

SINCE the publication of Dr. Erskine-Murray's excellent translation of Ruhmer's "Drahtlose Telephonie," there has appeared no complete exposition in any language, so far as the author is aware, of the important and growing field of radio telephony. The date of publication of Ruhmer's book was 1907, and in the succeeding decade advances have been made which render the methods there described obsolete.

Accordingly the author has accepted the considerable task of writing a full description of the radio telephonic methods of the present in the thought that he might thus contribute to progress in that valuable art. Wherever possible, sufficiently complete descriptions of the apparatus employed have been given to enable the "person skilled in the art" to duplicate the results and to judge their value. In this connection, the ranges of transmission and the output power of the radiophonic stations have been presented. It has been thought, too, that the arrangement of apparatus in the set was of value in that it indicated how the various designers had attacked the problem of making radiophonic apparatus readily usable.

It need hardly be stated that a good working knowledge of the methods and principles of modern radio telegraphy is necessary for the full appreciation of the material here set forth, though certain questions common to both types of communication have been here considered in detail.

The author desires to express his deep indebtedness for the valuable assistance in the form of information, illustrations, and diagrams which he has received from the following gentlemen and the organizations with which they are connected:

Atlantic Communication Company and Messrs. Boehme, Brockmann and Reuthe;

Compagnie Générale de Radiotélégraphie, and Lieutenants Colin and Jeance of the French Navy;

Lieutenant W. Ditcham of the British Army;

Mr. William Dubilier;

Federal Telegraph Company and Mr. Leonard F. Fuller;

General Electric Company and Messrs. E. F. W. Alexanderson, Albert W. Hull, and William C. White;

Mr. W. Hanscom;

Marconi Wireless Telegraph Company of America and Mr. Roy A. Weagant;

National Electric Signaling Company and Mr. John L. Hogan, Jr.;

Radio Telephone and Telegraph Company and Dr. Lee de Forest;

Mr. Bowden Washington;

Mr. Eitaro Yokoyama.

Through their kindness he has been enabled to present here much important information and previously unpublished material.

He also desires to commend highly the painstaking efforts of the editors of The Wireless Press to make this book measure up to the high aims of that organization.

ALFRED N. GOLDSMITH.

RADIO TELEPHONY.

CHAPTER I.

1. WORLD ASPECTS.

Before presenting to our readers the technical details of radio telephony, we shall discuss briefly the effects of this new type of communication on international affairs and world growth.

It is most difficult for a citizen of a modern state, beside whose breakfast table lies the printed sheet bearing the most recent news of widely distant happenings, to realize the elaborate and delicately adjusted mechanism which makes the entire earth his mental neighborhood. The labor of gathering accurate news, the transfer of these to the telegraph or telephone lines, the transmission of these across ocean or continent by the highly evolved radio telegraph or cable, and the huge task of editing, printing, and distributing them: all this shows but dimly in the final result. And yet, possibly the most fundamental difference between savagery and civilization and the most potent source of the latter is *communication*. The isolation of any modern state, the communication lines of which were irretrievably broken, would be truly tragic. The ties that would be broken would be not merely financial but in every field of human endeavor. Imagine a state which heard nothing of the politics, art, science, and literature of all the others. Picture the provincialism, the backward and undeveloped craving for the beautiful in art, the lack of co-ordinated scientific research and industrial development dependent thereon, and the childish literature which would result. A second "Dark Ages" of the mind and spirit would follow; and the citizens of the segregated state would be willing to pay almost any price for the restoration of communication. The evil effects of lack of communication on commerce need not be dwelt on; their magnitude and inevitability partake of the obvious. Commerce finds itself equally dependent upon rapid and reliable communication. It is exceptional that money in the form of the actual gold or silver is physically transferred from one country to another to settle debts. Payment is made by the transfer of credits between the countries or their merchants, and this transfer requires nothing more than the use of the radio or cable station for a few minutes. Only the small outstanding monthly or annual balance in favor of one or the other is physically conveyed between the merchants, and even this but rarely.

2. PERSONAL ASPECTS.

Aside from these larger aspects of communication, there are other advantages of communication which are priceless to the individual. The most obvious of these is the call for help in time of peril. We cannot gauge the value of a radio station on ship-board to the passenger or crew after collision or the breaking out of fire. The stringent laws of all nations relative to ship sets speak clearly for the opinion of the world. And marine law (and even naval law) have been altered by requiring the captain of the ship to remain directly and immediately responsible to his superiors on shore.

Modern business would, of course, be helpless except for the telegraph and telephone. Imagine our great companies in a world where all communication was by word of mouth, or by letter! The wheels of industry would turn but slowly when weighted down and clogged by slow and unreliable communication.

In the more personal matters of life, the literal extension of the personality by the telephone constitutes an inestimable privilege. The more pleasant social amenities become possible to all. Mere distance need no longer correspond to isolation, for, in effect, distance is completely bridged.

To summarize: in its larger aspects, COMMUNICATION IS THE LIFE-BLOOD OF CIVILIZATION, OF INTERNATIONAL GOOD WILL, AND OF PROGRESS.

To the individual, IT IS AT ONCE THE CLIMAX OF CONVENIENCE AND THE ULTIMATE EXTENSION OF PERSONALITY IN TIME AND SPACE.

3. USES OF RADIO TELEPHONY.

(a) The most natural use of radio telephony is from ship to ship and from ship to shore. Since it is the only means of telephonic communication possible under the circumstances, it does not need to compete with wire telephones or cables. By the use of amplifying relays at the receiving end (on shore), it will be possible to enable any person on the ship to communicate directly with persons on land, in part over the regular wire lines and in part by radio. The details of such communication will be explained in connection with "Radiophone Traffic." The great advantage of radio telephony over radio telegraphy on board ships is the direct personal contact between the persons corresponding and the resulting possibility of speedily settling the matters at issue, and (e.g., on freighters or tramp steamers) the freedom from the necessity of understanding the code. Of course, this last advantage is bound up with the simplification of ship radio telephone sets to the point where a skilled operator becomes unnecessary, the manipulation being simple and certain.

(b) A second important field for radio telephony is in trans-oceanic and trans-continental work. In the first of these, radio telephony is unique in meeting the requirements and is free from competition with submarine telephone cables. In the latter case, it would have to meet the competition of the long distance telephone lines. In each case communication between Subscriber A and Subscriber B would be through their wire lines to the nearest radio telephone high power station and thence automatically re-transmitted through an amplifying relay. This will be further explained in a later chapter.

(c) There are certain types of regions where radio communication is the only one possible of maintenance, e.g., in the arctic regions (because of snow and ice interference with wire lines), in densely wooded tropical regions (because of the enormous difficulty of maintaining a clear right of way through rapidly growing and luxuriant vegetation), in regions or across regions occupied in part by hostile savage tribes (who are addicted to the use of copper telegraph wire for ornament), and between islands and the mainland (where precipitous rocky coasts or swift currents injure or sweep away cables). In all of these cases, radio telephony offers its usual advantages and will no doubt come into increasing use.

(d) Between two moving trains or between moving trains and fixed land stations. Here, too, we are practically restricted to radio communication. The obvious advantages of such installations in times of storm, when wire lines are almost always incapacitated, has been shown by the experiences of the officials of the Delaware, Lackawanna & Western Railroad in times of blizzards. They have kept in touch with their

otherwise marooned trains, ·and have greatly simplified the problem
of resuming normal traffic schedules.   And even in fair weather, the
advantage of keeping all trains in touch with each other and the control
of train dispatching is obvious.   Occasional failures of the block system

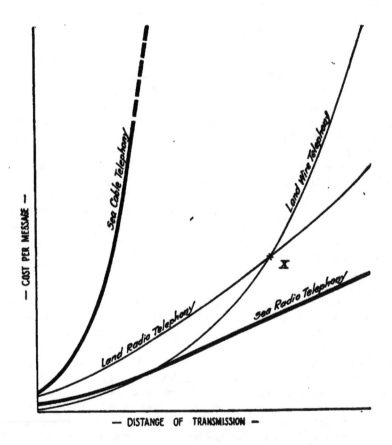

FIGURE 1—Comparison of cost; radio vs. wire telephony.

become far less dangerous, because it is possible to warn a train regardless of its position relative to the signals.   In foggy weather, this accurate
moment-to-moment information as to train positions is far from being
a drawback to the normally anxious passenger on certain railroads.

(e) There are a number of special applications of radio telephony
which have not as yet been developed to the point at which it is possible to

make any very definite statement as to their ultimate value. Among these are telephonic communication between various levels of a mine and the surface (which communication would greatly increase the chance of an early rescue in cases of cave-in, where wire lines are almost always broken), communication between government foresters, communication between aeroplanes or dirigibles and the ground, and communication between submarines and ship or shore.

4. RADIO VERSUS WIRE TELEPHONY.

It is very difficult, if not impossible, to institute a fair comparison between these fields at the present time. Radio telephony is so far from having reached an advanced stage of development, and is so seriously threatened on the research side by government control and naval or postal administration, that our conclusions are little better than guesses. However, certain broad considerations are fairly obvious and probable.

Let us consider Figure 1. Horizontally we have laid off on an arbitrary scale the distance over which telephone transmission is being carried on, the extreme distance covered by the chart being probably of the order of magnitude of 2,000 miles. Vertically, the cost for a three-minute toll message has been laid off, the extreme cost indicated being of the rough order of magnitude of $15 for three minutes. It is understood that these values may easily be as much as fifty per cent. or more in error.

(a) LAND TELEPHONY.

For short distances, there seems to be no question as to the superiority of wire transmission. The difficulty of preventing interference between a multiplicity of radio telephone stations, the first cost of even a low power radiophone station, the first cost of the transmitting and receiving antennas and ground, and the occasional skilled attendance required (at least, by present-day radiophones) render the idea of replacing the complex network of a city's wire telephone system by radiophones highly improbable. This feature is clearly shown by the lower portions of the fine line curves of Figure 1, wherein the influence of first cost on transmission over wire telephones and radiophones is qualitatively shown. There may be occasional exceptions to the curves shown; for example, in the case of very special types of service. Thus, it might be desirable for a military or police force to maintain radiophonic rather than wire line communication, for obvious reasons. But except when such special circumstances render radio communication imperative, the radiophone would seem to be at a disadvantage for short-range communication. As we gradually increase the range of communication, the circumstances may, however, alter. The vast expense of maintaining a two or three thousand mile long wire line, against sleet and snow, high wind, defective insulation, casual depredation, (and sometimes over-luxuriant vegetation) then come into consideration. If the wire line crosses one or more mountain chains, there are bound to be troublesome and weak points. Underground cables for wire telephony, except in the case of very high-grade and comparatively short-distance traffic, have not come into use because of their great cost. In addition, long telephone lines must be "loaded" electrically to prevent excessive speech distortion, and require the use of fairly elaborate two-way amplifiers at a number of points along the line. When it is considered that the cost of the line alone in the New York-San Francisco wire telephone transmission is in the neighborhood of two million dollars, and that this line must be constantly patrolled by hundreds of men, it will be seen that radio telephony may well come into consideration. That is to say, at some point (e.g., X in Figure 1), the radiophone may become more desirable than the wire telephone. There is no question that the distance of transmission corresponding to this point X depends directly on the extent to which strays can be eliminated in reception. It may safely be said that so long as radio telephony over long distances is dependent on absence of serious atmospheric disturbances, it will be handicapped thereby. With the advent of apparatus which markedly reduces stray intensity, wire line telephony over very considerable distances will be at a marked disadvantage. This will result in shifting the point X far to the left of the position indicated in Figure 1.

(b) OVERSEA TELEPHONY.

As soon as we consider telephony over water, we find a different state of affairs existing. It is questionable whether radio is not always less expensive than cable telephony in this case. Certain it is that over great stretches of water, radio telephony is at an enormous advantage because of the great cost of laying and maintaining the type of cable required for submarine telephony and also because radio communication over water is always accomplished with less power than for the equal distance over land. Consequently, we have tentatively indicated in Figure 1 the sea radio telephony curve as lying below the sea cable telephony curve throughout the length of each, and with the advantage of the former becoming specially marked for great distances. Of course, so far as long range oversea communication with ships is concerned, the radiophone has no rival.

Passing now to the technical aspects of radio telephony, we desire to make clear the scope of this book. It is not in the least intended to give every practical detail of construction of a "50 mile radiophone set," or indeed to go into many practical details of construction at all. The reason for this is two-fold. First of all, the limitations of space would prevent adequately treating all existing methods of radio telephony, even were all data available, and secondly, the cost to-day of building a reliably operative radiophone over any considerable distance is beyond the reach of most experimenters. In other words, the average amateur might just as well not attempt to construct such sets in the present state of the art. Furthermore, it is not possible for us here

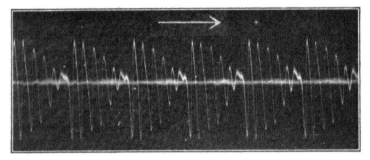

FIGURE 2—Oscillogram of vowel sounds "ah."

to give due credit to all those responsible for the historical development of each device described; nor to assign with any certainty patent rights in the apparatus mentioned. Present-day litigation and confusion as to patent rights would render such a course inappropriate on our part. We cannot even cover the entire field of radio telephony exhaustively. At best, we can only describe certain interesting and *operative* methods of radio telephony, assigning them to the manufacturer or designing engineer at present concerned with them, and giving proposed changes or improvements.

5. BROAD PROBLEMS INVOLVED IN RADIO TELEPHONY.

These problems are the following: (a) that of radiating energy at all for this purpose; (b) distortion of speech due to several causes; (c) the allied problem of amplification of speech at transmitter and receiver without distortion; (d) the obtaining of secrecy, and (e) the reduction of stray disturbances.

(a) RADIATION OF MODULATED ENERGY.

It first becomes incumbent on us to consider the nature of speech. In the back of the throat of the speaker a sort of membrane known as the "vocal cords" is set into more or less continuous vibration by the breath. The quality of the resulting sound is modified in at least two ways: by altering the shape of the mouth with the tongue or otherwise and thus causing a degree of selective resonance, and by actually starting or stopping the stream of sound as is done with the harsher consonants, e.g., the letter "d." The extreme complexity of the resultant sound vibration of the air is illustrated in the oscillogram of Figure 2.* This is a record of the current in a telephone line (and therefore approximately of the sound in the receiver) corresponding to the sustained vowel sound "ah" (as in "bah"), a clear-speaking man's voice being used for the test. The total time of the record is slightly over one-twentieth of a second. The basic vibration was of approximate frequency of 800 cycles per second and the chief modification thereof occurs with a frequency of 120 cycles per second. The great complexity of speech, even for the comparatively regular vowel sounds, is well illustrated. When the comparative crudity of radio telegraphy is considered, the difficulty of radio telephony becomes obvious. On the one hand, in telegraphy as nearly as possible complete and abrupt starting and stopping of the energy flow is required and this at no very rapid rate. In radio telephony, on the other hand, the outgoing flow of energy must be moulded and modulated with close approximation to the excessively complicated wave form of the speech vibrations. The difference in degree is not far from that between ruling a dot-and-dash line and making a dry-point etching of an autumn landscape.

Given, then, the complex vibrations which constitute speech, the problem of radiating the moulded energy arises. Of course, on a small and feeble scale the problem is solved in every-day conversation between two persons. This may be termed a species of "audio telephony," the frequency of the radiated air waves being those of the speech itself, i. e., of the order of 800 cycles per second. The same sort of solution might be attempted, using the electromagnetic "ether" waves of audio (i. e., audible) frequency to carry the telephone message. This solution is entirely unsatisfactory for a number of reasons. Firstly, the frequencies in speech vary considerably, and the radiating system (antenna) could not remain resonant to all these frequencies and their corresponding electromagnetic wave lengths. Secondly, the wave length would be excessively long, being 375,000 meters, or 230 miles, for the frequency of 800 cycles per second. This would require, for fairly effective radiation, an antenna of the length of say 10 to 20 miles, which is beyond the dreams of even the designers of the highest powered stations. Were an ordinary antenna about 300 feet (100 meters) high to be used, its radiation resistance at 800 cycles would be 0.0001 ohm, necessitating an antenna current of no less than 3,000 amperes to radiate even 1 kilowatt effectively. It is unpleasant to imagine the voltage at

* This unusually clear record the Author owes to the kindness of Mr. John B. Taylor.

the antenna top under these conditions; its value being not far from a million volts. Obviously, as a practical consideration, radio telephony by means of electromagnetic waves of the same frequency as that of speech vibrations is out of the question.

At this point the problem of radio telephony looks sufficiently hopeless; but fortunately an ingenious alternative (and a successful one) is available. Let the rippling curve of Figure 3 represent the sound vibrations corresponding to some spoken word. If this word was recorded on a vertical-cut phonograph record, a cross section of the groove of the

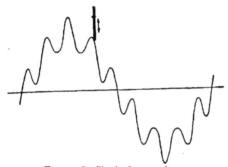

FIGURE 3—Typical wave-form.

record would show this curve as indicated. If a needle, indicated in the figure, were to move from left to right along the groove, and were pressed against the record it would also move up and down. If a diaphragm were fastened to the upper end of the needle, this diaphragm would set into motion the air near it, and the resulting sound vibrations would be an accurate reproduction of the original speech used in making the record. So far, we are on familiar enough ground.

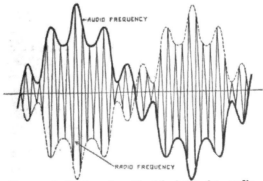

FIGURE 4—Basis of radio telephony by audio frequency modulation of radio frequency energy.

But suppose that we were suddenly to encounter a difficulty of the following kind: Imagine that it were not feasible to secure a large enough diaphragm at the top of the needle to set much air into motion. We might choose to use a small diaphragm vibrating *very rapidly* instead. In fact, we might arrange that this diaphragm vibrated so rapidly that its vibrations could not be heard at all, *but only the variation in their amplitude or width of swing.* Our phonograph record would now have to assume the curious appearance of the thin-line to-and-fro curve of Figure 4. This curve has been appropriately marked "radio frequency" in the figure, as distinguished from the heavy or envelope curve marked "audio frequency." The audio frequency curve is exactly the same as before, but it is replaced for radiating purposes by the *moulded or modulated radio frequency curve. The radio frequency curve* should strictly not have sharp peaks at the extreme of each alternation but should be a rounded "sine" curve. For clearness in the figure, it has been indicated as sharply peaked. Its frequency must be over 10,000 cycles per second, corresponding to inaudible "sound."

It may seem peculiar to speak of "hearing the variations in amplitude of a super-audible vibration," yet this is entirely possible. All we should need under the simplest conditions would be a "sound rectifier"; *i. e.,* a device which permitted only one-half of the radio frequency sound to reach the ear. This would correspond, in Figure 4, to admitting to the ear only those portions of the radio frequency vibration which lie above the middle line. Although the ear could not follow each of the myriad radio frequency impulses which it would thus receive, nevertheless the ear drum would receive inward pushes of an amplitude variation corresponding to the heavy-line audio frequency curve. Consequently the variations in the super-audible vibration would certainly be heard. The necessity for the "sound rectifier" is clear enough when we consider that without it extremely rapid impulses on the ear drum in opposite directions (corresponding to the entire radio frequency curve) would merely neutralize each other, causing no actual motion of the heavy ear drum. It is assumed that, though the ear drum can follow audio frequency vibrations readily enough, its inertia is so great that it could not follow the radio frequency vibrations to any appreciable extent. Hence the necessity for the "sound rectifier" producing mono-directional impulses of varying amplitude instead of bi-directional mutually neutralizing pushes-and-pulls of variable amplitude.

If we substitute for the explanation in the above imaginary acoustic case, the corresponding electrical case, we find that the explanation given holds equally. Since our antennas are too small electrically to radiate effectively audio frequency electromagnetic waves (as shown in an earlier paragraph), we are compelled to telephone by means of the variation of super-audible (that is, radio frequency) electromagnetic waves. In other words, the energy actually radiated from the station must resemble the "radio frequency" curve of Figure 4, and follow in its envelope curve (i.e. the audio frequency curve) the original sound vibrations.

The necessity for the crystal or valve *rectifier* (corresponding to the imaginary "sound rectifier" mentioned) is also evident if we substitute in the analogy already given the combination of telephone diaphragm and ear drum for the ear drum itself. Its function will be seen to be the furnishing of mono-directional mutually assisting electrical impulses which can push aside a heavy telephone diaphragm, which same diaphragm would hardly respond at all to the bi-directional mutually neutralizing unrectified impulses.

From the foregoing, we can draw one very important conclusion. The radio frequency used in radio telephony must be quite inaudible and completely steady, and many times higher than the audio frequency voice vibrations. Otherwise we should hear in the receivers a continuous, high, and piercing tone corresponding to the ever-present radio frequency, which shrill tone would naturally be an objectionable interference with the conversation. Furthermore, the accurate reproduction of the delicate overtones in the voice, which are of fairly high frequency themselves, is dependent on having many radio frequency cycles available for the moulding process, so that the envelope curve will be very faithfully followed.

It is to be noted that a second method of radio telephony exists, which might be termed "modulation by change of frequency (or wave length)." Instead of altering the amplitude of the radiated waves in accordance with the envelope speech curve, we might systematically increase and diminish the radiated frequency in proportion to the envelope curve. For example, while normally radiating at 50,000 cycles per second (6,000 meters wave length), we might alter the frequency say to 48,000 cycles at points corresponding to the peaks in the audio frequency curve, to 49,000 meters for points corresponding to half-way between peak and zero in the audio frequency curve, and so on. At the receiving station, the response in the detector circuit would then be proportional (or nearly so) to the speech curve in view of the tuning and detuning effects which would occur in the receiver as the rapidly varying frequency was received. This method permits keeping appreciably full load on the radio frequency generator at all times.

It is the view of the writer that any such method is objectionable in that it distributes the radiated energy over a considerable range of wave lengths, thereby increasing the liability to interference with other stations. Furthermore, stray reduction will probably require the reception of a single sharply defined frequency.

A third alternative method exists for radio telephony, this being a combination of the first two. That is, both the amplitude and the frequency of the radiated waves are varied in accordance with the audio frequency curve. This method, rather than the second, has been occasionally used; but it suffers from the same defects as the second method and has no great advantages over the first.

(b) CAUSES OF SPEECH DISTORTION IN RADIO TELEPHONY.

In radio telephony we are, of course, vitally concerned in preserving faithfully the exact quality of speech from the speaker to the ear of the person receiving the message. That is, the wave form of the original sound (as shown in Figure 3 and also in the dotted outline in Figure 6) must be in no way distorted in transmission and reception. This requires considerable care in the various stages of the process of radio telephony, as will appear from the following review of the causes of distortion and their remedies.

To begin with, there is a type of distortion which may be termed "fly wheel" or inertia distortion. It arises in a fashion which can be made clear from a mechanical analogy. If we have a fly wheel in rapid rotation, there is a marked and well-known tendency of the wheel to maintain a constant speed because of the large amount of energy stored in its rotating mass. If we attempt to change the speed of the wheel very greatly in an exceedingly brief time, we shall meet with almost insuperable opposition. The "inertia reaction" of the wheel will be very marked. If, on the other hand, we attempt to change the speed slightly in a considerable longer time, we shall find the task a much easier one. In other words, the fly wheel resists, markedly, rapidly recurrent changes in its speed of rotation but permits, fairly well, slow changes in its speed. Of course, the same effect exists with any mass. If we attempt to start and stop a heavily loaded freight car a thousand times per second, huge forces will be called into play if the to-and-fro swing of the car is appreciable. If we attempt to start and stop the same car only once per second, the opposition will be but a thousandth as great.

The application of this principle of inertia reaction to the telephone transmitter diaphragm and the telephone receiver diaphragm is not far to seek. It is clear that there will be much more difficulty in getting the telephone diaphragm to respond proportionately to the higher overtones of the human voice than to the lower pitched components. This is one reason why a high-pitched voice generally fails to get over a telephone line with complete satisfaction to the listener. We may say, then, that the telephone transmitter diaphragm smooths out the high overtones of the voice, with the result that crisp, clear enunciation is in part lost. The obvious remedy is to have light transmitter diaphragms without much pressure on them from the carbon grains behind the diaphragm. We are much limited in the design of a transmitter by other considerations, so that it is generally necessary to use the normal transmitter. As regards the receiver diaphragm, there have been made a number of attempts to use thin small sheets for this purpose, and it has been noted that they respond much more readily to the present-day 500-cycle spark signals than did the older, heavier diaphragm receivers. On the other hand, it is desirable to avoid receivers in which the diaphragm is markedly resonant to any frequency within the normal range of speech, else this frequency will be relatively exaggerated out of all proportion to its actual magnitude. The result will be an extremely annoying "howling" whenever the resonant frequency occurs in the speech.

There is another possibility of a sort of "inertia" distortion due to the fact that the radio frequency generator (arc, bulb, alternator, etc.) will not supply sudden violent changes in output. Consequently, a similar smoothing down of the higher overtones will occur unless the radio frequency generator is without "electrical inertia" (that is, has small inductance) and also is operated by a powerful generator.

It is well at this point to indicate clearly what is meant by "electrical inertia." The behavior of an ordinary inductance when an alternating electromotive force (voltage) of various frequencies is applied at its terminals is as follows: The current which passes through the inductance is inversely proportional to the frequency; that is, the inductance acts like an electrical mass and does not permit the ready passage through it of the higher frequency currents. It is to be noted further that the effect of a capacity is exactly the opposite in that it exaggerates the passage of currents of higher frequency while relatively retarding those of lower frequency. This is the basis of ordinary tuning, which is nothing more than balancing the choking action of an inductance at a given frequency by the opposite effect of a capacity.

It is clear enough then that high inductance in any of the circuits in which speech currents flow will produce an objectionable smoothing out of the overtones in speech so that the speech will become "drummy" through the exaggeration of the basic tones. An excess of capacity in any speech circuit will exaggerate the overtones, and the speech will become "squeaky." A little practice in telephony soon enables the skilled observer to tell which type of distortion he is getting.

A smoothing-out effect, causing drummy speech, is also obtained when a highly persistent antenna is coupled to the sustained wave generator or when a highly persistent receiver secondary is coupled to the antenna. This is due to the fact, first shown by Bjerknes, that with loose coupling a persistent secondary will not follow the sudden variations in the primary oscillations except with a time lag and diminution in the abruptness of change. It is this effect which has led Poulsen and others to use entirely aperiodic secondaries in their radiophone receiving sets. In this way, the persistency distortion in coupling is avoided, though at the cost of loudness of signal and selectivity, *i. e.*, freedom from interference from other stations. These matters will be further considered under receiver design.

A fairly prolific source of speech distortion, or rather speech destruction, is that curiously imperfect instrument, the carbon microphone as used in the ordinary telephone transmitter. This has been shown by de Forest, who states that when speech from an ordinary transmitter is very greatly amplified, it is found to be fairly teeming with crackling and hissing sounds caused by small local arcs or mechanical disturbance effects in the microphone. This is not astonishing when the nature of the variable-resistance carbon contact is considered. The alternative suggested by de Forest, namely the use of an ordinary Bell receiver *as a transmitter of the induction type* is feasible only if one is willing to amplify greatly the extremely small output of such a transmitter. On the whole, the investigator will in general be bound to use carbon transmitters with small currents so as to avoid "packing," hissing, and other disturbances.

If iron cores are used in the coils of transmitting circuit (e. g., in magnetic amplifiers, or speech-controlled frequency doublers), a further distortion will arise because of the "saturation" effect in the iron. That

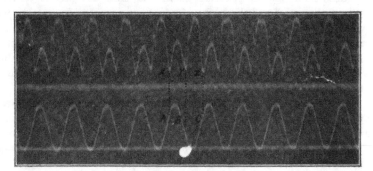

FIGURE 5—Oscillogram showing distortion when range of linear proportionality is exceeded.

is, iron core coils do not have a constant inductance at all, and the change in their inductance is particularly marked when heavy currents are passed through the coils thus saturating the iron. An interesting example of the distortion produced in a magnetic amplifier is taken from Mr. E. F.

W. Alexanderson's paper in the April, 1916 "Proceedings of the Institute
of Radio Engineers." The lower curve of this figure shows an oscillogram of the current (amplified speech current) passing into the magnetic
controlling amplifier. The upper curve shows resulting voltage of the
controlled radio frequency alternator. The distortion is seen to occur
between points X and Y (corresponding to A and B in the lower curve)
but not between points Y and Z (corresponding to B and C in the lower
curve). That is, the distortion occurs for the high current values in the
iron-core amplifier control winding between A and B. Of course, such
effects are avoided by working the iron on low field density so that the
flux is at all times proportional to the magnetizing current and the control
range is not exceeded. This can be accomplished, though sometimes at
the cost of great amplification and large output.

It will be noticed that in discussing the saturated-iron distortion, we
have encountered a case of non-linear amplification, and resulting distortion. Non-linearity of amplification is sufficiently common and important to warrant detailed consideration.

(c) NON-LINEAR AMPLIFICATION AND SPEECH DISTORTION.

Let us consider again an ordinary phonograph record of speech,
and let us suppose that the record in question is to be "amplified"

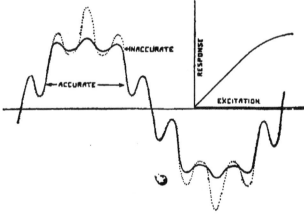

FIGURE 6—Non-linear distortion.

mechanically. That is, we wish to produce a record similar in all
respects except that the up-and-down ripples in the groove are to be
accurately magnified to a definite extent in their vertical dimension but
their length is to remain unchanged. Such a record would produce

a sound of much greater loudness but with the pitch or frequency unchanged. (We are here referring to a vertical cut record, of the cylinder type.) This amplifying procedure would be quite satisfactory if the mechanism that cut the new record always followed the original record accurately, and faithfully multiplied every vertical dimension by the same amount. Then, in Figure 6, the new record would have the cross-section of the dotted line in the figure. If, however, the amplifying mechanism magnified accurately only for displacements near the axial line but responded disproportionately feebly for portions of the curve lying at considerable distances from the axial line, we should get the type of distortion shown in the full line of Figure 6. It will be seen that the overtones are blurred at the upper portion of the curve, which is accordingly labeled "inaccurate." In the lower portion of the curve, where linear proportionality is obtained, the curve remains accurate. The whole matter is shown in a different way in the right hand diagram of Figure 6, where input of the amplifier or "excitation" is plotted

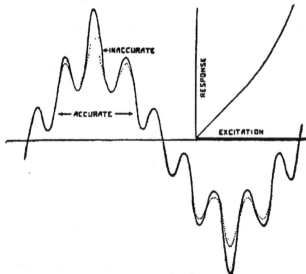

FIGURE 7—Non-linear distortion.

horizontally against output or "response" vertically. It will be seen that, for the accurate portion of the wave, the response-excitation curve is a straight line, hence the name "linear amplification." Up further it flattens out, somewhat like an iron saturation curve and it is here that the distortion occurs. Speech of this type would generally be called "drummy."

In Figure 7, the reverse type of error, leading to what is usually termed "squeaky" speech, is pictured. It will be seen that the amplification is linear for low excitations, and that consequently the lower portions of the wave near the axial line will be accurately amplified. On the other hand, the greater excitations produce a disproportionately great response, and the overtones are exaggerated near the peak of the wave. The result is a high tinny quality to the speech.

In Figure 8 is shown a sort of combination effect of these two, which is not uncommon. It consists of a disproportionately feeble response for small excitations, a proportional response for moderate excitations, and a disproportionately feeble response for great excitations. The resulting wave is, as will be seen, accurate only for moderate values, but flattened as to overtones near the axis and far from the axis. This would be badly blurred or "muffled" speech.

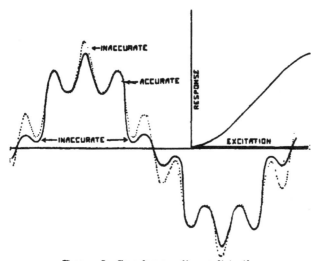

FIGURE 8—Complex non-linear distortion.

It is quite clear that we should use linear control systems in the radiophone transmitter and linear amplifying systems in the radiophone receiver. With the magnetic amplifier for transmission, this implies lower iron field-densities, and with the audion receiving amplifier it implies working below the saturation current point.

(d) SECRECY OF COMMUNICATION IN RADIO TELEPHONY.

If a frank expert were to be asked whether "complete" secrecy could be obtained now with radio telephony, he would be compelled to answer in the negative. If he were of a cynical turn of mind, he might add that secrecy was no more obtainable in radio telephony now than in wire telephony or any other form of communication, a remark which recent revelations as to the comparative prevalence of official "wire-tapping" would more than justify. Of course, any wire telephone line can be tapped, and with remarkable ease under most conditions. At one time, the telephone companies judged it necessary to maintain wire communication from New York to Boston over one line and return communication from Boston to New York over another line traversing an entirely different route from the first. In this way, even the adroit interloper would hardly be likely to tap more than half a conversation. As a matter of fact, the drastic expedient suggested was not adopted since it was unnecessary then and will probably continue to be so. A combination of severe laws against tapping, and an efficient corps of radio inspectors would practically solve the problem, at least in communities no more law-defying than those of the United States.

As an illustration of the ease of tapping an ordinary (non-twisted, though regularly transposed) telephone line, it may be mentioned that there is much used abroad a simple secondary coil, which, when placed suitably near the line, picks up ordinary conversation without any visible physical connection, permanent injury, or other trace. Even the effect on the transmission would be practically infinitesimal.

It is to be expected, on the other hand, that when radio telephony becomes commercial and widespread, we shall have stringent laws against "listening-in" on commercial wave-lengths, and these laws will be adequately enforced. By the use of a number of modified waves or by other technical methods under development at present, unauthorized "listening-in" will become exceedingly difficult, and possible of attainment only by persons of expert ability. Such persons, however, are easily known and can be supervised in their activities; much as is now the case with excessively skillful engravers of bank notes. In fact, systems can be imagined whereby "listening-in" would be futile unless the listener had a code combination whereby the peculiar material sent could be automatically re-converted into clear speech. This indicates a possibility of complete secrecy.

In short, while secrecy in radio telephony involves more inspection than for wire telephony, it can be brought to the same or even a greater degree of certainty by technical and legal measures.

(e) STRAY INTERFERENCE IN RADIO TELEPHONY.

One of the most serious outstanding problems in radio communication is the elimination of atmospheric disturbances and stray electromagnetic waves. To begin with, under normal summer conditions, particularly in the tropics, the effect of strays is practically to prevent stations from working at all, part of the time. Aside from the six-to-one to ten-to-one ratio of signal strength in favor of winter time, the strays complicate the problem of reliable transmission vastly. As a result, most radio stations work with a "factor of danger" rather than the normal engineering "factor of safety." Whereas an engineer will normally design, for example, a bridge so as to stand five or ten times the maximum load which it will be called on to bear, (that is, with a factor of safety of ten), the radio engineer is unable to guarantee transmission over great distances, particularly in the tropics, without the use of excessive and commercially unprofitable amounts of power. The result is that a compromise is always made between absolutely reliable service (and no profits) and moderately reliable service on a financially feasible basis.

If however, ninety per cent. of the strays could be eliminated in reception, the effect would be virtually to increase by ten times the power of every transmitting station and to render communication entirely reliable even where it had been previously fairly uncertain. It has, indeed, been estimated with probable correctness that in the absence of strays (or their practical elimination) communication from Germany to the United States could be carried on with about ten kilowatts in the antenna, or even less. When it is considered that at present a power of two hundred kilowatts in the antenna at Nauen is not really always sufficient, the importance of stray elimination is made increasingly evident.

Radio telephony has one great advantage over radio telegraphy in the matter of stray elimination. It is well known that speech can be carried on, after a fashion, even under very serious difficulties; for example, in extremely noisy localities. The ease in understanding speech under such conditions is due particularly to our lifelong practice, since it is rather unusual (in cities at least) to carry on speech under conditions of even approximate silence. Then, too, there is what may be termed the "assistance of context." By this is meant the ability of the average listener in "filling in" lost words in a conversation by judging what word, placed in the gap, would give sense to the entire sentence. This assistance is much greater than is usually believed, as can be easily shown by the common experience in listening to names over a telephone. Whereas ordinary conversation is carried on over normal telephone lines without any particular difficulty, the moment names or figures (that is, material lacking assisting context) are given, great difficulty is experienced and the percentage of errors rises markedly.

There is no doubt, therefore, that understanding a telephone conversation through comparatively heavy strays is a simpler achievement than taking down telegraphic signals under the same conditions.

We shall return in some detail to the methods used in stray elimination, or reduction, in connection with receiving systems (page 220).

CHAPTER II.

It will be seen from the previous treatment of the subject of radio telephony that a complete one-way installation comprises a generator of practically sustained waves at the transmitting station, a means for controlling or modulating the output thereof, an antenna and ground system for radiating a portion of the modulated energy; and, at the receiving station, an antenna and ground system, and a radio receiver with or without suitable amplifying devices.

It is proposed to consider first the various types of sustained wave generators which may be used in radio telephony.

6. SUSTAINED WAVE GENERATORS.

(a) ARCS.

For the sake of completeness, we shall give here a description of the theory of the arc and its historical development from one of our earlier papers:

"The simplest generator of radio frequency oscillations of considerable power is the Duddell-Poulsen arc. In Figure 9 is shown the arrangement used by Duddell. G is a direct current generator, R' is the resistance intended to control the arc current, and L' a choke coil intended to keep the alternating current out of the generator and also to steady the supply voltage. K (for the Duddell arc) has solid carbon electrodes. L, C, and R are inductance, capacity, and resistance inserted in the arc shunt circuit. Their values should be carefully chosen.

"If the arc be lit, it is found that an alternating current appears in the shunt circuit, and if the frequency of this current is within the limits of audibility, a pure singing tone will be heard."

21

The arc differs from ordinary conductors in one essential respect. If we divide the potential difference (or voltage) at the terminals of an ordinary metallic conductor by the current flowing through the con-

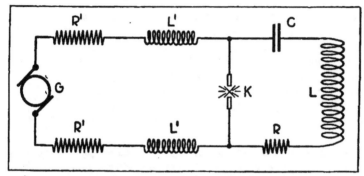

FIGURE 9—Typical arc circuit.

ductor, the quotient is found to be a *constant* quantity called the resistance of the conductor. This is the case regardless of the values of voltage and current (at least, until the conductor becomes heated by the passage of excessive current). In the arc, the quotient of voltage divided by current is by no means constant. In fact, for high voltages the arc resistance is a large quantity and very little current passes

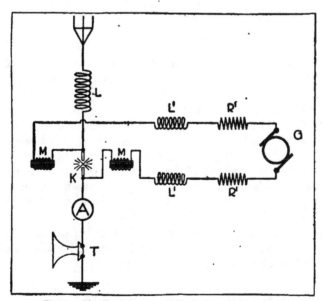

FIGURE 10—Typical arc radiophone transmitter.

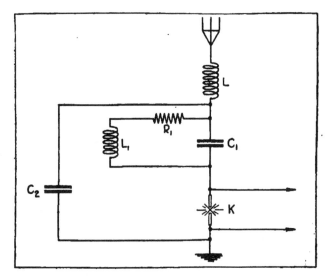

FIGURE 11—Fuller's method of increasing arc efficiency.

FIGURE 12—Berliner-Poulsen arc for portable
stations.

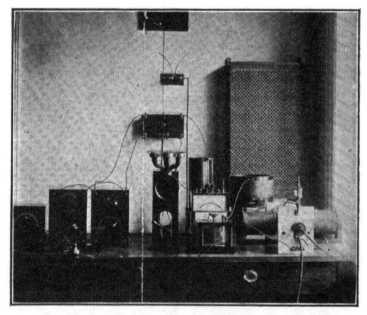

FIGURE 13—Danish Poulsen arc radiophone transmitter.

through the arc under such voltages. For moderate voltages, the arc resistance is much less, and moderate currents pass. For low voltages, the arc resistance becomes exceedingly small and the arc current tends

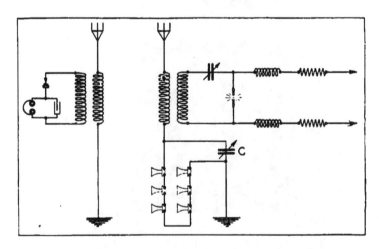

FIGURE 14—Poulsen radiophone transmitter and receiver.

to increase indefinitely; that is, the arc is unstable and tends to become a short-circuit. We are forced then to the conclusion, that a small *increase* in the voltage at the terminals of an arc causes a small *decrease* in the resultant current; and consequently we sometimes speak of the "negative resistance" of an arc as distinguished from the ordinary or positive and current-limiting resistance of metallic conductors.

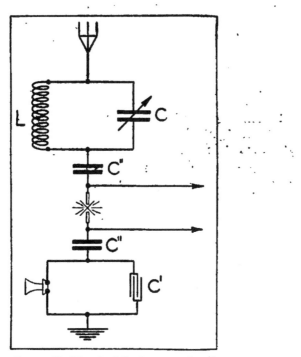

FIGURE 15—Fly-wheel Poulsen arc circuit
for radio telephony.

"The theory of the action of the singing arc is the following: When the condenser and the inductance in the shunt circuit are connected to the arc, the condenser begins to accumulate a charge, and therefore robs the arc of a part of its current, since the supply current is kept appreciably constant by the presence in the supply leads of the choke coils L'. If the current through the arc decreases, it is clear from the foregoing considerations that the voltage at its terminals must increase. Consequently, as long as the charging of the condenser continues, the arc voltage will rise. As soon as the condenser is fully charged, the arc voltage becomes stationary. Then the condenser begins to discharge itself through the arc, thereby increasing the arc current and diminishing the voltage. The shunt circuit being a true periodic or oscillatory circuit, the discharge of the condenser will continue past the point of zero current, and there will occur an actual reversal of current. Thus the condenser becomes charged in the negative direction until the arc voltage falls so far that the supply voltage of the direct current generator causes a reversal of the whole action. The cycle is then repeated and with a frequency related to a certain extent to that of the natural oscillations of the shunt circuit. The mode of vibration which takes place in the arc is thus closely analogous to the action in an organ pipe of the reed type.''

In 1903, Poulsen raised the arc to the status of a practically operative generator of radio frequency energy in considerable quantity by the following changes: placing the entire arc in an atmosphere of hydrogen or a hydrocarbon vapor (e.g., alcohol or gasoline), using a carbon electrode for the negative side and a copper anode water-cooled for the positive side, rotating the carbon electrode slowly by motor drive, and placing an intense deflecting magnetic field transverse to the arc. Except for certain constructional and electrical details, this is the Poulsen arc of to-day.

In Figure 10 is shown a typical arc radiophone station. The arc K is shown in the magnetic field due to the coils M. A is an ammeter for measuring the antenna current, and T is a heavy-current transmitter, usually of the carbon microphone type. The control methods which may be used with arcs other than that shown will be considered under another heading. A modern improvement in arc transmitters, and one which results in a great increase in overall efficiency of the arc, is that shown in Figure 11. This is due to Mr. L. F. Fuller, and the inventor states that very marked increases in output result when it is used. It consists first: in placing in shunt with the arc and the antenna series condenser C_1 the condenser C_2 and second: in placing around the series condenser C_1 an inductance L_1 and a resistance R_1. The chief function of the condenser C_2 is to act as a bypass for the radio frequency current, thereby avoiding passing through the arc the entire antenna current as well as the direct supply cur-

FIGURE 16 — Lorenz-Poulsen arc for radio telephony.

rent. The circuit $L_1R_1C_1$ is tuned nearly to the frequency of the antenna current and thus acts as an absorbing circuit for such currents. It will therefore have the function of a powerful choke circuit for the arc and will assist the condenser C_2 in its action.

Before discussing the actual results obtained by Poulsen arc radiophone transmitters, we shall show the types of construction of such arcs in sizes rated from 250 watts to 100 kilowatts. The first of these, shown in Figure 12, is a small portable military station made by the Telephone Manufacturing Corporation of Vienna. In this case, the carbon is rotated at intervals by hand, and the alcohol feed into the arc chamber is accomplished automatically by the vaporization of the alcohol by the heat of the arc.

As early as 1906, Poulsen established radiophonic communication over a distance of about 600 feet (200 m.) using antennas only 15 feet (5 m.) high. In 1907, with the equipment shown in Figure 13, communication was established between Esbjerg and Lyngby, a distance of 170 miles (270 km.). The antenna height was 200 feet (60 m.), the wave-length 1,200 meters, the arc supply power 900 watts, and the antenna power 300 watts. Later, phonograph music was heard in Berlin after transmission from Lyngby a distance of about 300 miles (500 km.). In Figure 13, the arc is shown to the extreme right, and the multiple microphone transmitter (six microphones in series) in the middle of the figure. The arc was in its own primary circuit in this case, and coupled inductively to the antenna. At the left, the inductively coupled receiving set is shown. The secondary circuit was made aperiodic by placing

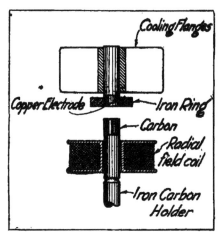

FIGURE 17—Construction of Lorenz-Poulsen arc for radio telephony.

the detector directly in series with the secondary coupling inductance and without any secondary tuning condenser. The reasons for this type of receiver will be discussed under another heading. In Figure 14 are illustrated the arrangements used. It will be noted that the microphones are shunted by the condenser C, thereby making the transmission partly one by change of wave-length as well as by change in antenna energy.

When small antennas of high intrinsic decrement are used, an arrangement known as a "fly-wheel circuit" may be employed. This is shown in Figure 15. The circuit LC is inserted in the antenna, L being large in comparison with the antenna inductance. The wave-length of the radiated energy will consequently be approximately that of the circuit LC. In this way, energy may be stored in the highly undamped circuit LC and gradually radiated. The two condensers C'' are not essential to the operation and serve only to keep the direct current supply leads from conductive connection to antenna and ground thereby avoiding the possibility of serious high voltage shocks when touching the antenna. In the case shown, the microphone-shunting condenser, C' has a value between 0.05 and 0.20 microfarad.

In Figure 16 is given a photograph of an arc rated at about 100 watts output, and built by the C. Lorenz Company of Berlin specially for radio telephony. Its construction is shown in Figure 17. The carbon holder is of iron, and forms the open core of a circular multi-layer coil which produces a moderately strong magnetic field passing directly upward through the carbon. The field then spreads outward to the upper iron ring, passing radially through the arc in so doing. In consequence of the presence of this field, the arc will slowly rotate around the edge of the carbon, thereby causing even wear. The copper electrode is held within the iron ring, and provided with massive cooling flanges. In Figure 16, the alcohol sight-feed cup is seen at the top, and the vertical cooling flanges below and to the right. The insulator between the upper and lower electrodes is a heavy ring with flat faces, made of plaster with asbestos facings. The clamping screws are also visible, as are the two poppet safety valves which relieve the excessive pressure resulting when the arc is first lit and the mixture of alcohol vapor and air explodes. The lower electrode holder and the surrounding coil are just below the middle of the illustration.

One of the defects of these small arcs is the necessity for adjusting the arc length occasionally as the carbon burns away. To overcome this, the Lorenz Company has built a self-regulating arc provided with a mechanism somewhat like the magnetic length-control of an ordinary street lighting arc. The device is supposed to be very effective in practice.

FIGURE 18—3 k. w. Berliner-Poulsen arc.

A somewhat larger type of arc of 3 k.w. input made by the Telephone Manufacturing Corporation (formerly J. Berliner), of Vienna, is illustrated in Figure 18. It will be noted that provisions are made here for an extremely intense magnetic transverse field. While this is of advantage in increasing the available radio frequency output, it tends to cause a certain degree of irregularity in the output with a resultant crackling noise or hissing in the received speech. This last defect may prove extremely serious, so that magnetic fields on arcs used for radio telephony must be employed with caution, and with associated circuits and outputs which minimize arc unsteadiness. In Figure 18, the small motor which rotates the carbon electrode is seen in the lower central portion, and there are also shown the sections of the magnetic field whereby the field strength may be conveniently varied. The heavy insulation surrounding the push button used for striking the arc at the beginning of operation is a necessary adjunct since disagreeable burns are easily sustained when using high power arcs carelessly. An arc of somewhat smaller output than that shown is guaranteed by the makers for telegraphy over 125 miles (200 km.) when using portable masts, but for only about 12 or 15 miles (25 km.) for radio telephony.

A somewhat larger type of arc, built by the Danish Poulsen Company (Det Kontinental Syndikat), is shown in front view in Figure 19. The massive field magnet coils, the driving motor for the carbon holder, and the arc striking and adjusting knobs are visible. It is to be noted that all arcs giving any considerable output have air core choke coils in the feed circuit, since the distributed capacity of iron core coils gives rise to the possibility of injurious resonance phenomena inside the coils and may permit radio frequency currents to pass.

FIGURE 19—Continental Syndicate Poulsen arc.

An arc made by the Berliner Company of Vienna, and having an input of 10 to 25 kilowatts is illustrated in Figure 20. As will be seen, it is provided with an automatic ignition device, a device for the indication of arc length or wear of the carbon electrode, a mixing chamber of glass for the gas used in the arc, and a complete water cooling system for the field magnet coils as well as for the arc.

Some interesting information relative to Poulsen arc radiophones of the type shown in Figure 21 is given by Captain Anderle. Figure 21 is a ship station of a complete type, being used for either telegraphy or telephony. The arc, which is normally rated at about 8 kilowatts input. is shown at the left. For telephony, it is used at reduced power, inasmuch as the multiple microphone transmitter at the right would be

FIGURE 20—Berliner 25 k. w. Poulsen arc.

incapable of modulating the full output. About 3.5 amperes (and never over 4) are passed through the transmitters, which are placed directly in the ground lead of the antenna. With masts 150 feet (45 m.) high, distances of from 30 to 60 miles (50 to 100 km.), are covered over flat country. The speaker is warned to speak distinctly but not too loudly, with the mouth held near the transmitter. It is recommended to tap the microphones occasionally, or to have alternative sets so that overheating of one set does not occur.

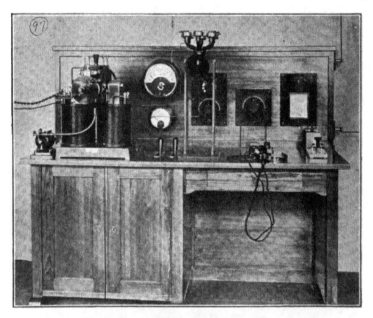

FIGURE 21—Berliner-Poulsen arc ship radiophone station.

FIGURE 22—Berliner-Poulsen 3 k. w. arc radiophone station.

In Figure 22 is shown an unusually complete set of the Poulsen arc type built by the Berliner Company of Vienna. This set is adapted at the same time to ordinary arc telegraphy, multi-tone arc telegraphy, and radio telephony. The arc is of 3 k.w. input, being the same as that given in Figure 18. The telegraphic range of this set is given as 375 miles (600 km.). The receiving set and test buzzer are mounted on the right hand portion of the long table; the arc and key at the left center; the relay key and transfer switches are to the left of the arc near the variable transmitting coupling and the multi-tone control keyboard. In the extreme left foreground is the large microphone transmitter, to be described hereafter when control systems are considered.

FIGURE 23—Federal Telegraph Company 60 k. w. Poulsen arc at Tuckerton.

Although the newer, high power arcs are not yet employed for radio telephony because of the great difficulty in modulating the output, nevertheless they form a possible direction of radio telephonic development. Accordingly. we show in Figure 23 an arc of 60 k.w. input made by the Federal Telegraph Company. this corresponding to 500 volts and 120 amperes. It is this arc which has carried a portion of the trans-Atlantic traffic from Tuckerton, New Jersey, to Hanover, Germany, a distance of 4,000 miles (6.500 km.). In this case, the antenna current was 120 amperes, the arc standing considerable overload. We also show. in Figure 24, a 100 k.w. arc made by the same Company, and used for communication between the United States Naval Radio Station at Darien, Panama Canal Zone and Washington, a distance of 1,900 miles (3,000 km.). It will be seen that both these arcs are very sturdily designed and provided with an unusually rugged and elaborate water-cooling system.

FIGURE 24—Federal Telegraph Company 100 k. w. Poulsen arc at Darien.

The Federal Telegraph Company carried on radiophone experiments from 1910 through 1912 between the stations at San Francisco, Stockton, Sacramento, and Los Angeles (all in California). The distance between any two of the first three stations mentioned is 90 miles (140 km.) and between the first and the last station mentioned 355 miles (570 km.). Speech between San Francisco, Stockton, and Sacramento was clear at all times, but between these points and Los Angeles it was weak and indistinct. The antenna heights were 300 feet (94 m.)

Before leaving the subject of the Poulsen arc, it is of interest to give detailed accounts of just what has been accomplished by these methods in addition to the achievements already mentioned.

A remarkable series of experiments were made by Q. Majorana in 1908. The results obtained are best described in the words of Majorana himself:

"The first research was conducted in the Istituto Superiore del Telegrafia in Rome. The antenna was 78 feet (24 m.) high and had four wires. For two years, I have been conducting experiments between this station and a government naval station at Monte Mario, 3 miles (5 km.) away. The antenna at the latter station had also four wires and was about 175 feet (50 m.) high. An ammeter in the antenna at the former station showed under normal conditions of working a reading of about 1.2 amperes. At Monte Mario, using the thermo-electric (crystal) detector, a current of 15 micro-amperes was obtained. Words spoken in Rome could be heard at Monte Mario by the use of a Marconi magnetic detector, but could be heard very much more loudly and clearly by the use of the former detector.

"Because of these results, the Naval Bureau provided a second research station at Porto d'Anzio, 25 miles (56 km.) from Monte Mario, with a four-wire antenna 145 feet (45 m.) high. On the 13th of August, 1908, the experiment was tried, and this showed that with a current strength of 3.5 amperes in the antenna at Monte Mario any words spoken in Rome could be very distinctly heard at Anzio.

"Hereupon the Naval Bureau ordered that these researches should be carried on over longer ranges. The torpedo boat 'Lanciere' was accordingly put at my disposal, and arrived on the 13th of November at the Island of Ponza, about 75 miles (120 km.) from Monte Mario. On this island there is a station for radio telegraphy, with an antenna of four wires about 200 feet (60 m.) high. Using the same receiving apparatus as had been employed at Rome, words spoken in Rome could be heard at Ponza with greater loudness even than at Anzio; the vibrations of the telephone diaphragm could be heard at a distance of 10 or 15 feet (3 or 4 m.). The superiority of these results is to be ascribed to the better location of the station at Ponza.

"On the 14th of November, the 'Lanciere' landed at Maddalena in Sardinia. The nearby station at Becco di Vela, which is similar to that at Ponza, was then used. The station is about 170 miles (270 km.) from Rome in an air line. On that day, at 12 o'clock, attempts to communicate with Rome were repeated and again gave excellent results.

The voice at Monte Mario was distinctly audible, and the strength of the speech was not less than it is in the ordinary wire telephone in use in the city. We can, therefore, state that over this range a practically workable radio telephonic service can be provided.

"Finally, I desired to find the utmost range of the radio-telephonic apparatus at my disposal. On the 1st of December, the 'Lanciere' arrived at Trapani, in Sicily, where further attempts were made, using the radio-telegraphic station at Monte San Giuliano. This station resembles that at Ponza, and is 270 miles (420 km.) in an air line from Rome. It took quite some effort to secure sharp tuning here, partly because of considerable interference from a neighboring station, but finally the spoken word from Rome could be heard, even though it was faint and not easy to understand. The intensity in this case was barely sufficient for the trained ear to read. We were here at the limit of the range. This was proved more clearly on the following day. At Forte Spurio is a station which is about as far from Rome as that on San Giuliano, but less favorably situated. I went to Forte Spurio and found that the words sent from Rome could not be heard there.

"The utmost range of my system was by no means reached in these experiment, for the hydraulic microphone was not used to a point even approaching its full capacity." (Majorana used a Poulsen arc generator. but modulated the antenna energy by means of a special hydraulic transmitter which will be described under control systems, page 152.)

"I desire to mention one important point in these experiments. After several trials, it was positively shown that the quality of the word was not altered, even at distance of 250 miles (400 km.). That is, the articulation was clear and the fine inflections of the voice were preserved. This is because all the various frequencies contained in the speech suffer the same weakening for equal distances, so that there is no distortion of the speech. With the ordinary telephone lines, on the other hand, the propagation depends largely on the acoustic period; in radio telephony, the period of the electro-magnetic radio frequency oscillations is of the greatest importance."

Experiments were carried on at the end of June, 1909 between the large Poulsen stations in Denmark at Lyngby and Esbjerg, the distance between these stations being 170 miles (280 km.). The Egner-Holmström heavy current microphone, to be described later (page 144), was used directly in the transmitting antenna. Such microphones can carry 10 to 15 amperes, but it was shown that this current was unnecessary for the range in question. In fact, with an antenna current of 6 amperes properly modulated, communication of a very good and clear sort ("sehr gut and deutlich" according to the experimenters) was established.

An arc system of radio telephony distinguished by simplicity rather than by efficiency or perfect reliability in practice has been developed by the Telefunken Company, though it has been superseded by their radio frequency alternator-frequency changer methods to be described later.

The arcs used by the Telefunken Company were burned either six in series on 220 volts direct current, twelve in series on 440 volts, or twenty-four in series on 880 volts. They burned practically in the open air. The lower carbon electrode rested in a depression in the base of a large, hollow, copper cylinder filled with water, which cylinder formed the other electrode. The water naturally served for cooling, and the carbon dioxide formed by the slow combustion of the carbon remained partially in the depression mentioned, and prevented the further and free access of air to the arc. No magnetic field was used with the arcs in question, and the efficiency was low. With an energy consumption of 6 kilowatts for 24 arcs in series, only about 10 per cent. of the available energy was converted into the radio frequency form. However, the carbon electrodes which were 3.5 cm. (1.4 inches) in diameter burned nearly 200 hours for each half inch of length.

FIGURE 25—Telefunken Company series arc
radiophone transmitter.

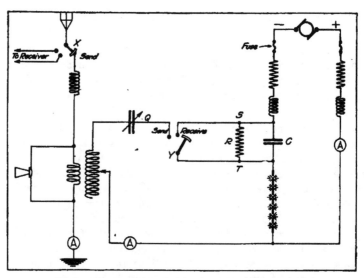

FIGURE 26—Telefunken series arc radio transmitter.

The arcs were arranged as illustrated in Figure 25. It will be seen that all six could be struck at once by the right-hand handle, and that the length of each arc could be adjusted individually by a separate adjustment screw (not shown in the illustration). The actual wiring of the set is shown in Figure 26, and presents some valuable features. To begin with, there is a switch, X, which not only transfers the antenna connection from the transmitter to the receiver, but short-circuits the receiver while transmission is going on, by the use of auxiliary contacts, not shown. The switch, Y, connects together the points, Q and S, while

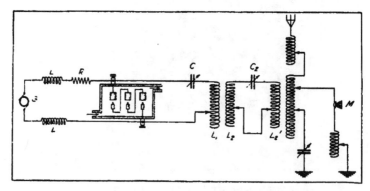

FIGURE 27—Colin and Jeance radiophone transmitter.

sending is going on and the arc is oscillating. While receiving is going
on, the oscillations are stopped by opening the connection between Q and
S. At the same time, the resistance, R, becomes operative in holding
down the direct arc current. During transmission the alternating current generated by the arc passes through the condenser, C, while only the
direct current passes through R. In this ingenious way, the arc current
is prevented from rising markedly when the oscillations cease, which is
otherwise the case. The microphone is seen to be connected across the antenna tuning inductance, which also serves for coupling. Consequently,
the microphone has the triple purpose of diminishing the coupling,
shortening the radiated wave-length, and diminishing the antenna current by dissipating a portion of the available energy.

On November 15, 1907, using the apparatus just described, radiophone speech was transmitted from Berlin to Rheinsberg, a distance of
about 45 miles (75 km.), the mast heights being 85 feet (26 m.), and
the input power 440 volts and 5 amperes.

In 1908, a system of radio telephony developed by Lieutenants
V. Colin and M. Jeance of the French Navy was first thoroughly tried
out.* There were used three arcs in series supplied with 600 volts, the
three being regulated simultaneously (and in later models automatically).
An oscillatory circuit is shunted around the arc, and coupled to an
intermediate circuit, to which the antenna is coupled in turn. The
positive electrodes are heavy copper cylinders with cooling (usually by
an interior stream of kerosene), and the negative electrodes are carbon
rods of extremely small diameter (1 or 2 mm., i. e., 0.04 to 0.08 inch),
the arcs taking place in an atmosphere of some hydrocarbon such as
illuminating gas, acetylene, gasoline, alcohol, heavy oils, etc.

Under these conditions, the positive electrodes are not attacked at
all, and the negative (carbon) electrodes merely increase slowly and
regularly in length because of the deposition thereon of a fine layer
of carbon from the hydrocarbon atmosphere. Consequently, the arc does
not tend to wander about the electrodes as is usual.

In order to ensure purity of the radiated wave and freedom from
overtones (which are apt to prove troublesome, particularly for arcs in
powerful magnetic fields, or arcs from which excessive energy is being
drawn), the arc oscillating circuit is coupled to the antenna by means of
an intermediate tuned circuit which, in turn, is inductively coupled
to the antenna.

The microphone transmitters are of special design and contain
no combustible material, the grain carbon being placed in cavities cut

* For much of the information here given, the Author is indebted to the "Bulletin de
la Société Internationale des Electriciens," for July, 1909.

into sheets of marble or slate. The vibrating diaphragm is held at a suitable distance from the carbon support by a metal washer.

The actual circuit employed during some tests in 1914 is shown in Figure 27. Here G is a 650 volt direct current generator, which furnished in the tests in question 4.2 amperes (that is, 2.73 k.w.). This current passed through the choke coils L and L', and the large regulating resistance, R, to the arcs which are shown schematically in cross section. The relative size of the electrodes and the method of admitting the hydrocarbon gas atmosphere are indicated. The potential difference across the arcs in this case was 350 volts, and consequently the energy consumption in the arc was 1.47 k.w., leaving 1.26 k.w. to be absorbed in the circuit of the generator, G and probably mainly in the resistance R. The intermediate circuit. $L_2C_2L_2'$, was very slightly damped, the capacity, C_2, being large. It will be noted that the microphone, M, is shunted across a portion of the antenna coupling and tuning inductance, being itself in series with an inductance. It will thus have the triple function of altering the coupling to the intermediate circuit, altering the radiated wavelength, and absorbing intermittently a portion of the available radio frequency energy. The main antenna current was 3.2 amperes at a wavelength of 985 meters, and the current through the microphones was 0.5 ampere. Nine microphones in series were employed, and two sets were provided for alternate use to avoid overheating.

FIGURE 28—Motor generator control panel of Compagnie Générale de Radiotélégraphie-Colin-Jeance Radiophone Transmitter.

The arc carbons in these tests were 1.5 mm. (0.06 inch) in diameter, and the arc took place in an atmosphere of acetylene (from calcium carbide and water) mixed readily in proper proportions with hydrogen (from calcium hydride and water). Under these conditions, the arcs were not burnt away; in fact, the carbons increased slightly in length with operation. Independent arc length regulation was provided for each arc, but was not found necessary.

Flat spirals of copper strip were employed in the various circuits, and either air variable condensers or glass fixed condensers. An auxiliary tone circuit shunted around the arc was provided (not shown in the figure), whereby musical note telegraphy could be easily accomplished. Since the total terminal arc voltage dropped from 350 to about 150 when not transmitting, an auxiliary resistance was provided in the

Figure 29—Colin and Jeance series enclosed arcs, automatic regulator, and control apparatus.

supply circuit which was automatically shunted into circuit whenever reception was begun.

With the equipment shown, communication was maintained between Paris and Mettray, a distance of 200 kilometers (125 miles).

In Figures 28, 29, 30, and 31 are shown the various assembled portions of a modern complete set of this type, as manufactured by the Compagnie Générale de Radiotélégraphie.   The panel of Figure 28 supports the motor and generator switches, measuring instruments, and control rheostats.   The three arcs and their enclosing chamber together with the arrangement for their gas supply are illustrated in the upper part of Figure 29.   An automatic regulator for the arcs is mounted directly in front of them.   In the lower portion of the table are mounted the supply circuit choke coils and resistances. The means for tuning the primary, intermediate, and secondary (or antenna) circuits are provided in the cabinet shown in Figure 30.   The hot wire ammeter at the top is in the intermediate circuit.   The two couplings between the pairs of circuits are controlled by the projecting handles.   Figure 31 illustrates the operator's table.   The measuring instruments are the antenna ammeter, the microphone, shunt circuit ammeter, and a voltmeter across the arcs.   The resistance to the right is in the microphone circuit.   The two microphone mouthpieces and reversed horns and the change-over switch between sets of microphones are also on the back panel.   On the table top are the antenna switch, a change-over switch from telegraphy to telephony,

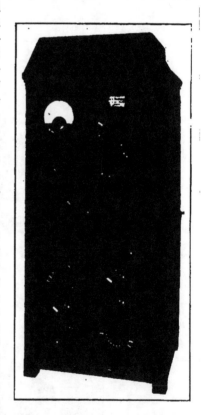

FIGURE 30—Colin and Jeance primary, intermediate, and secondary control panel.

the sending key, the enclosed detectors, and a complete receiving apparatus. This last is of normal type, having inductive coupling between the antenna and secondary circuits, a tuned secondary, and crystal detector.

FIGURE 31—Colin and Jeance transmitting and
receiving operator's table for radio telephony.

CHAPTER III.

(b) RADIO-FREQUENT SPARKS.

It has occurred to a number of investigators that practically sustained radiation could be secured in an antenna by using spark transmitters, but having these transmitters so arranged that the extremely high frequency of the sparks (above the limits of audibility) would render the usual "spark tone" inaudible. If, then, the antenna energy were modulated by a microphone or otherwise, radio telephony would become possible. To specify in further detail, imagine a special form of spark gap and associated circuit so arranged that discharges occurred more or less regularly across the gap at an average frequency of, say, 50,000 sparks per second. If the circuit in which these sparks occurred were connected inductively to an antenna, there would be produced in the antenna practically sustained radiation, susceptible to suitable telephone modulation by a microphone transmitter or otherwise.

In Figure 32 is given a graphic delineation of the effects. It will be noticed that highly damped oscillations occur rather irregularly in the primary circuit, and that each of these short oscillation groups starts a decadent wave train which has still a large current amplitude when the succeeding spark takes place. Inasmuch as the sparks follow each other so frequently and since the antenna circuit damping is low, the effect at the distant receiver would be appreciably that of sustained radiation at the transmitter, and particularly is this the case since the changes in antenna radiation occur above audio frequency. Most of the radio-frequent spark transmitters for radio telephony operate in the fashion indicated, but there is a second special case, which has certain interesting

44

features. It is illustrated in Figure 33, and occurs with the Chaffee "arc" (which is really a spark phenomenon). To begin with, in this case the spark gap has such excessively high intrinsic damping that the spark discharges in the primary circuit tend to be aperiodic. (The structure of the Chaffee arc is described on page 56.)

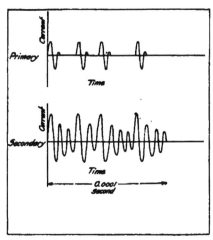

FIGURE 32—Irregular radio-frequent spark
excitation of antenna.

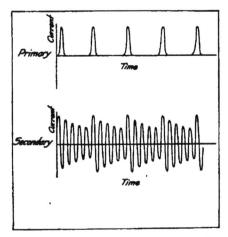

FIGURE 33—Regular radio-frequent impulse
excitation of antenna.

The tendency toward aperiodicity just mentioned is enhanced by Chaffee in that he couples the secondary circuit very closely to the primary, thereby obtaining a "quenching" action through the secondary reaction on the primary. In addition, the direct current feed circuit of the arc and the coupling to the energy-absorbing secondary are so arranged that the spark frequency is an integral fraction (e. g., one-half. one-third, one-fourth, etc.) of the frequency of the oscillations in the secondary circuit. Thereby it occurs that the successive discharges come at just the right time to be in phase with the secondary (or antenna) oscillations, and not at random (with possible interference) as is the case for the conditions illustrated in Figure 32. In Figure 34 are shown oscillograms of the actual phenomena.* Chaffee uses the term "inverse charge frequency" for the ratio between the

FIGURE 34—Primary and secondary current with Chaffee gap. (Inverse charge frequency = 3)

radio frequency in the secondary circuit and the spark frequency in the primary circuit. The inverse spark frequency is a whole number for the Chaffee arc.

In general, spark methods of radio telephony are open to very serious objections. Unless the sparks not only follow with great regularity but also have nearly equal current amplitudes (neither of which

FIGURE 35—Ruhmer moving wire arc for radio telephony.

conditions are easily fulfilled, particularly in steady operation), there will be produced in the receivers of the distant station an annoying hissing sound, which will interfere seriously with clear articulation in the speech. This accounts, naturally, for the frequently poor quality of spark radiophone transmitters.

Nevertheless, many investigations have been carried on in these directions, and in some cases with marked success, and these will now be considered.

In the fall of 1900, working with a special interrupter enabling him to obtain as many as 10,000 regular sparks per second, Fessenden succeeded in transmitting speech over 1 mile (1.6 km.), but the quality was poor and there was much noise. By 1903 better speech was

* By courtesy of Mr. Bowden Washington.

obtained, though the extra noise was still present. It does not appear that Fessenden developed this method further, although he describes a special rotary gap (with 40 per cent. platinum-iridium studs) operated on 5,000 volts direct current and arranged to give 20,000 sparks per second by the successive charging and discharging of a condenser.

One of the earlier workers with radio-frequent spark systems was Ernst Ruhmer. Ruhmer used as his gap terminals two moving metallic wires which passed over water cooled prismatic surfaces at the sparking point. The apparatus is shown diagrammatically in Figure 35, which shows clearly the reels on which the moving wire is wound and the

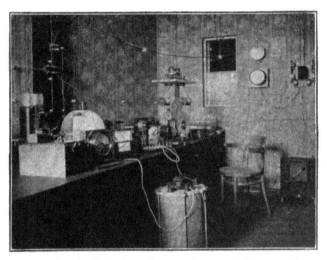

FIGURE 36—Ruhmer's radiophone transmitter.

water-cooled prismatic gap guides. The paramount advantages of Ruhmer's arrangement are that a fresh and clean surface is constantly presented for the arc, that excellent cooling (and consequently quenching) is obtained, and that the arc length should remain quite constant. Ruhmer's apparatus is shown complete in a hitherto unpublished photograph* in Figure 36. The arc mechanism is shown on the table near the extreme right. The reels from and to which the wire passes, the driving motor, and the two cup-shaped containers for cooling water on the top of the apparatus just over the arc are visible. These cups surmount the gap bearings. At the right end of the table are the controlling rheostats and lampboard resistances which regulated the supply of high voltage direct current. On the table can be seen the micro-

---

*Which photograph, together with a number of others shown herein, I owe to the courtesy of Mr. William Dubilier.

phone transmitter, antenna and closed circuit ammeters, coupling and inductance coils, and in the foreground a wave meter.

Another early system (1911) of radio telephony with what the inventor, Mr. William Dubilier, called a "quenched arc" (really a radio-frequent spark) transmitter, is indicated in outline in Figure 37. The arc is indicated at A, and is fed with moderately high tension direct current. Shunted around the arc is the oscillatory circuit, $C\ L_1$, which is opened automatically by a simple switch during reception. The oscillatory circuit, or primary, is coupled to the antenna by means of the inductive coupling between L_1 and L_2, and also by means of a capacitive coupling through the condensers C_1 and C_2. Shunted across the condenser C_2 is the telephone relay R of special construction to be described here-

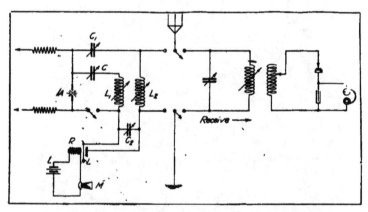

FIGURE 37—Dubilier radiophone transmitter and receiver.

after. It is practically a heavy current microphone transmitter coupled to an ordinary receiver electromagnet, the electromagnet in question being energized from the master microphone M. Mr. Dubilier has pointed out that the terminals of an ordinary telephone line may be substituted for the local microphone connections at L, L, thus causing the incoming energy from the telephone line to control the heavy current telephone relay, R, and enabling direct communications from any usual land line telephone station to a ship at sea.

The receiving set is indicated at the right of Figure 37, and is an inductively-coupled, aperiodic secondary, crystal detector receiver of fairly conventional design.

In Figure 38 is illustrated a complete radiophone station of this type. The box at the left of the table contains most of the transmitting equipment. On the right rear corner of the top of the box is the multiple contact commutator for changing from transmitting to receiving. This commutator performs all the necessary functions indicated by the switches in Figure 37. On the top of the box is a moderately heavy current, multiple microphone transmitter, consisting of a number of transmitters (7) in series. At the right of the box is mounted the special gap or discharger. It consists of one heavy, well cooled metal electrode and one small uncooled electrode. The antenna inductance and coupler is shown in the middle of the table and, at the right of the table, the receiving set.

A later and improved type of set is shown in Figure 39, which is

FIGURE 38—Dubilier radiophone transmitter and receiver.

the entire transmitter self-contained. The antenna commutating switch has been somewhat improved, and the antenna ammeter is mounted on the top of the apparatus box. The details of the gap, including the horizontal fins for air cooling, are clearly shown. This particular set has an input of about 3 k.w. and has enabled radio telephony 250 miles (400 km.) on one occasion. The containing box is only 14 inches on a side 35 cm.). The tilted side at the left of the box has mounted on it one of the spiral coils of the antenna coupling, so that merely changing the angle of inclination of the exterior tilted side varies the antenna-to-primary coupling.

1

FIGURE 39—Complete Dubilier radiophone
transmitter.

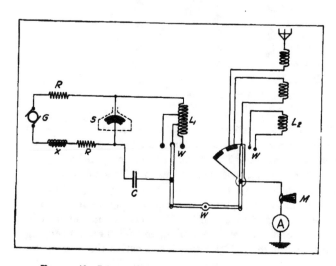

FIGURE 40—Lorenz Company radiophone transmitter.

FIGURE 41—Lorenz Company aeroplane
"multitone" transmitter.

The C. Lorenz Company, of Berlin, has developed through its engineer, Dr. H. Rein, a system known as the "multitone" system. Though primarily intended for low and medium power, variable tone, radio telegraph transmitters it has also been employed in radio telephony. The circuit diagram of the set is given in Figure 40. Here G is a moderately high voltage direct current generator, R and X are feed circuit resistances

FIGURE 42—Interior of Scheller "multi-
tone" gap.

and choke coils, W is a wave changing switch, which, after a preliminary and final adjustment of the taps on L_1 and L_2 enables choosing instantaneously any one of three wave-lengths. The microphone is placed in the antenna as indicated. Dr. Rein pointed out (as had also, and independently, Dr. Seibt) that the resistance of the microphone for best modulation should be equal to the total resistance of the remainder of the antenna circuit. This would imply that one-half the available energy would be consumed in the transmitter microphones, a rule that obviously limits the available modulated output of sets of this type.

A small aeroplane set of this type (intended for telegraphy, however), is shown in Figure 41. The gap, which is the most interesting portion of the set, is seen in the left corner. It consists of two nearly concentric spherical segments, one fitting within the other. The construction is given by Figure 42, which is the dis-assembled gap. The discharge takes place in an atmosphere of alcohol vapor, the alcohol being supplied by the top sight-feed cup. The gap was devised by

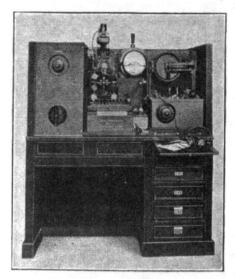

FIGURE 43—Lorenz Company "multitone" ship set.

Scheller. A complete ship station of this type is given in Figure 43, and a semi-high-power station in Figure 44. This last has gaps for high tension, low frequency alternating current, the gaps being assembled in groups of six in series. In radio telephony, Rein states that in general carbon grain microphones having resistances between 4 and 10 ohms were used. If necessary, these were coupled to the antenna through a suitable transformer, or otherwise, in such fashion that the equivalent resistance they interposed in the antenna circuit was equal to the remaining antenna resistance.

Another system of radio telegraphy that has been adapted to radio telephony in quite a similar manner to the latest mentioned is that due to E. von Lepel. The circuit used is identical with that of Figure 40 in some cases, though in the recent 2 k.w. sets the circuit shown in Figure 42 is used. This is analogous in action to that shown in Figure 11 (but with L, R_1, and C_1 omitted), and operates in the manner there explained, at least to some extent. The spark gap shunt circuit L_1C_1 is tuned to nearly the same frequency as the plain antenna circuit.

The Lepel gap consists of a plane bronze negative electrode separ-

FIGURE 44—Lorenz Company "multitone" semi-high power set.

ated from a plane copper positive electrode by a thin sheet of "bond" paper, say 0.002 inch (0.05 mm.) thick. The center of the paper sheet is perforated, and when approximately 500 or 600 volts direct current is applied between the electrodes, the discharge bridges the gap. It then continues rambling outward, slowly burning up the paper sheet, in a spiral path starting at the center and ending at the edges. This action is probably due to the deflecting action of the electrostatic field between the plates on the discharge current. Another circuit used by von Lepel is shown in Figure 45.

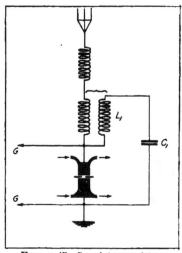

FIGURE 45—Lepel transmitter.

The Lepel gap is usually shunted by an audio frequency oscillating or "tone" circuit, when used for telegraphy. When used for telephony, however, the gap is unshunted and a very rapid succession of discharges occur, each setting up its train of waves in the antenna, as indicated in Figure 32. In the receiver there is then heard a faint hissing sound. By inserting a microphone in the antenna, this hissing is drowned out by the speech, and telephony becomes possible. A Lepel radio telegraph set (at Harfleur, France), is shown in Figure 46. The spark gap, which is water-cooled, is seen just to the left of the large coupler. It can easily be dis-assembled for cleaning and replacement of the paper separator.

FIGURE 46—Lepel station at Harfleur.

Continuing our discussion of radio telephony by means of radio-frequent spark transmitters, we consider next a system developed by Dr. E. Leon Chaffee in conjunction with Professor George W. Pierce. This system will be found to be unique in certain respects.

The wiring diagram of the transmitter is shown in its essentials in Figure 47, and presents no unusual features. The direct current generator supplies 500 volts (and from 0.3 to 0.8 ampere; i. e., from 150 to 400 watts) per gap. The resistance provided in the supply circuit is made in two parts, in series, one roughly variable in considerable steps and the other smoothly and continuously variable. This is desirable, since the operation of the gap, though steady, depends on a proper choice of the current, this current partly determining the

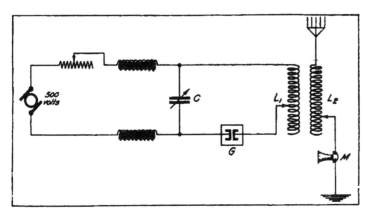

FIGURE 47—Chaffee radiophone transmitter.

inverse charge frequency. The phenomenon of an inverse charge frequency (that is, a whole-number ratio between the secondary oscillation frequency and the primary impulse frequency) has been treated above, and is illustrated in Figure 33. It constitutes a distinctive feature of the Chaffee gap, and depends on the intrinsically great damping in the gap.

The primary condenser.C need not be.a high tension condenser with the usual low power sets, and generally has a·value in the neighborhood of 0.009 microfarad. The coupling between L_1 and L_2 is close. Ordinarily, the microphone M is an ordinary Bell transmitter, though Chaffee has stated that this type of microphone deteriorates somewhat under radio frequency currents of one ampere or more.

The cross section of a Chaffee gap, constructed by Messrs. Cutting and Washington (under patent license from Dr. Chaffee) is shown in Figure 48. The gap consists of plugs of aluminum and copper respectively, one or two square centimeters (or roughly two or four-tenths of

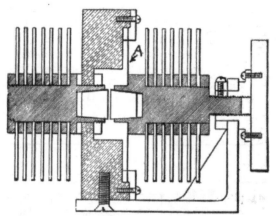

FIGURE 48—Cross section of Chaffee gap. Designed by Cutting and Washington.

a square inch) in area, larger dimensions being undesirable in the stationary forms of the gap. Originally the gap was run in an atmosphere of moist hydrogen; but hydrogen being difficult to obtain in ordinary practice, it was found by Cutting and Washington that alcohol vapor could be substituted provided it was distilled into the gap, by the gap heat, from a wick entering the bottom of the gap chamber. The form

FIGURE 49—Chaffee gap. Designed by Cutting and Washington.

of gap shown is made air-tight by the use of the flexible phosphor bronze diaphragm, A, which is held in place against a soft rubber gasket by a brass ring. Such a diaphragm permits the necessary movement required in adjustment of the gap electrode separation. The external appearance of the gap with its adjusting handle and cooling fins is given in Figure 49. For larger powers, a still later modification of the gap is used wherein the discharges pass between a rapidly rotating aluminum disc and a stationary copper plate, in hydrocarbon vapor. High efficiency (up to 60 or 70 per cent.) can be obtained with these last gaps.

The discharge begins when the switch is closed, provided the distance between the electrodes is not over 0.1 mm. (0.004 inch). It is a noiseless and fixed arc of a vivid violet or purple color. Occasionally

FIGURE 50—Aeroplane set with Chaffee gap. Designed by Cutting and Washington.

it moves to a fresh point on the electrodes. The explanation of the extreme quenching action lies, according to Chaffee, in "the practically instantaneous re-establishment of the high initial gap resistance when the current becomes zero, due probably to the formation of an insulating oxid film on the aluminum; the high cathode drop of the anode metal; and the absorption of energy by the secondary, although rectification usually takes place without this aid." The best operating gap lengths are from 0.04 to 0.09 mm. (0.0016 to 0.0036 inch).

The primary discharge is a half loop of current, and, as correctly indicated in Figure 33, is not half a sine wave. Its duration does *not* depend on the primary supply current, which latter affects only the time between successive primary discharges. The time between successive primary discharges is also dependent on the primary capacity, since the charging phenomena connected therewith largely determine the successive breakdowns of the gap. For an inverse charge frequency of 2 or 3, the secondary oscillations differ only imperceptibly from truly sustained oscillation, as is evidenced by the interesting fact that when received on a normal beat receiver, a clear musical beat tone is obtained.

It is worthy of note that even with this absolutely aperiodic primary discharge, a definite relation between the primary period and the secondary period is required for maximum secondary response. This relation is, however, far from being one of even approximate equality being, in fact, a ratio of 1.71 for primary period divided by secondary period.

FIGURE 51—Front view, 0.25 k.w. Chaffee
gap set. Designed by Cutting and
Washington.

The radio frequency output per gap is about 50 watts, and the efficiency is given as between 30 and 40 per cent. Two or three gaps may be operated in series on 500 volts, and four gaps on 1,000 volts. The actual voltage drop across the individual gap is about 150 volts.

The Chaffee apparatus as developed for commercial work by Cutting and Washington is illustrated in Figures 50, 51, and 52. The first of these is a 150-watt aeroplane set, with the special gap in the center. The primary condenser is behind the gap, and the primary-to-antenna coupler is mounted to the left. In the latter two figures, a somewhat larger set is depicted. Here two gaps in series are used, and a variometer type of coupling. Telegraphic communication was maintained with one of these sets 78 miles (125 km.) with 1.5 amperes in the antenna at 480 meters wave-length. It should be noted that, in marked contrast to almost all sustained wave generators, the Chaffee arc drops but slightly in output at very short wave-lengths.

It has been pointed out elsewhere by the Author that a marked tendency exists in radio development toward having all stations operate with sustained radiation. This tendency is much to be encouraged because of the remarkable possibilities in the direction of selectivity with beat reception at the short wave-lengths. While beat reception is not particularly suited to radiophone work, it is to be hoped that ship and

FIGURE 52—Side view, 0.25 k.w. Chaffee
gap set. Designed by Cutting and
Washington.

small shore stations, and all amateur stations will at least employ sustained wave generators. If this is done, the Chaffee arc would seem to be a suitable device, and has marked possibilities.

In the radiophone experiments described by Chaffee, great simplicity of apparatus was achieved. The regular tests were carried on over a distance of one mile (1.6 km.). A single gap was used with from 0.2 to 0.5 ampere through it. The voltages at all portions of the set in the station were comparatively low, say under 1,000 volts. It is stated that when the receiving station was properly tuned, only a slight hum or hiss was heard in the receivers, which was tuned out, if desired, and in any case drowned by the voice. The articulation was very good, and communication was maintained for hours without losing a word or making any adjustments.

The speech was heard at a distance of 40 miles (64 km.), but it is believed that this distance is by no means the limit of the system, even when only one gap is used.

Mr. Washington has informed the Author that using two gaps and an antenna current of 2.7 amperes modulated by a water-cooled

FIGURE 53—Ditcham radiophone transmitter and receiver.

transmitter, music from a phonograph was clearly distinguishable on shipboard at a distance of 110 miles (180 km.).

Another system of somewhat similar characteristics was developed by Lieutenant W. T. Ditcham in 1912, and presents some features of interest. There was used a gap, the cathode of which was aluminum, hard copper, or bronze, the anode copper or steel, each electrode about 1 cm. (0.4 inch) in diameter, and the discharge taking place in an atmosphere of carbon dioxid under pressure. Four such gaps were used in series, at a voltage of 1,000 and a current of 1.5 amperes. The capacity in the primary oscillating circuit was 0.012 microfarad.

The description of the apparatus given by the inventor makes it clear that he was aware of the advantage of securing an integral inverse charge frequency, and attempted to secure this advantage in designing the apparatus.

The antenna fundamental was 460 meters, and its capacity 0.0007 microfarad. It was normally used at 550 meters with an antenna current of 8 amperes. The antenna was lower than desirable, and probably had only small true radiation resistance. The normal distance of communication was from Letchworth to Northampton, a distance of 55 km. (34 miles). However, signals have been received 175 km. (110 miles) over land. In reception, a crystal detector (namely, Pickard's silicon-arsenic combination) was used.

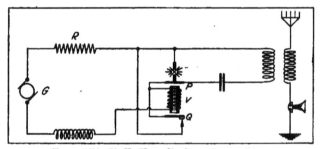

FIGURE 54—T. Y. K. radiophone transmitter.

It is interesting to note that the maximum radiation was attained in the system when the primary was tuned to 830 meters and the antenna to 550 meters, a ratio of 1.51 between them. This ratio is not far from 1.71, the value found by Chaffee for most efficient operation. Coupling to antenna as high as 40 per cent. is used.

We are indebted to Lieutenant Ditcham for important previously unpublished data on the operation of these gaps. With hard copper or bronze electrodes in carbon dioxid under pressure, the gas apparently had two functions: (a) cooling by expansion; (b) the formation of a hard crystalline film on the electrodes. This film permitted actual contact of the electrodes without "short-circuiting" or arcing. When the film was once formed, the gas could be shut off, and the spark would continue active for five or ten minutes before an arc started.

The entire transmitter is given by Figure 53. On the top shelf are mounted the four series gaps. On the shelf below are seen tuning inductances and a relay, while on the bottom shelf is mounted the receiver and a call-bell system. This last consisted of a Brown telephone relay fed from the crystal detector and, in its turn, supplying the current for a moving coil relay of no great sensitiveness. A long musical dash is sent for calling, the pitch being regulable by variation of the speed of the rotary make-and-break device ("chopper") which is inserted in the coupling between the closed and antenna circuits. A

FIGURE 55—Front view of T. Y. K. radiophone transmitter and receiver.

selective method of calling, permitting ringing any one of a number of stations within a given zone was experimented with, but no details are available as to its success in operation.

A system of radio-frequent spark telephony has been devised by Messrs. Wichi Torikata, E. Tokoyama, and M. Kitamura. The spark or arc terminals in this system are composed of magnetite (oxid of iron) and brass. Other alternatives are aluminum silicon, ferro-silicon, carborundum, or boron against minerals such as graphite, meteorite, iron or copper pyrites, bornite, molybdenite, marcasite, or others. Usually the electrodes are of small surface, this being regarded as essential by the inventors. The power supplied per gap is 500 volts and 0.2 ampere. A capacity of approximately 0.05 microfarad is used in the primary oscillating circuit. About 1 ampere is modulated in the antenna by the microphone, and the every-day range is given as 10 to 15 miles (25 km.). Ordinary crystal detector reception is employed.

The wiring diagram of the apparatus is given in Figure 54. It will be seen that the starting device is of an unusual nature. It seems that a high-resistance film forms on the surface of the electrodes, as in

FIGURE 56—Equilibrator and spark gaps of
T. Y. K. radiophone transmitter.

Lieutenant Ditcham's system, and it is necessary in consequence to have some means of obtaining a momentary high voltage to break down this surface film, and start the discharge. This is accomplished by having a steady current flowing normally (before oscillations are desired) in the inductance V as indicated, this current being quickly broken at Q when it has once fairly started. The gap electrodes being in contact, the high inductive voltage breaks down the surface film, and the armature P draws the electrodes apart and serves as a sort of automatic arc length regulator thereafter.

Figure 55 illustrates the transmitter proper and receiver. A normal heavy-current microphone transmitter is used (mounted at the top in front of the equilibrator). The primary oscillating circuit,

FIGURE 58—100-to-500-volt direct current
coils for direct current supply circuit
of T. Y. K. radiophone transmitter.

capacity control switch is directly below the microphone. The receiver
is mounted in the lower case, together with the ''sending-to-receiving''
switch. The crystal detector is enclosed in a metal housing, the door
of which appears at the lower left side of the receiving apparatus case.
A usual test buzzer and normal tuning and coupling coil switches are
provided. The equilibrator is shown in Figure 56, with the alternative
spark gaps (aluminum-brass or aluminum-magnetite), at the lower left
corner. A small lamp with cover is mounted at the rear to indicate

FIGURE 58—100-to-500 volt direct current
rotary converter of T. Y. K.
radiophone transmitter.

antenna current. The lamp resistance and choke coil box for the high voltage generator, supply circuit to the gap appear in Figure 57. The 100 volt (and 2.7 ampere) to 500 volt (and 0.2 ampere) rotary converter is illustrated in Figure 58.

In June, 1913, there were established eight land stations of this type in Japan and seven stations were installed on board ship. It is stated that commercial service was initiated at this early date.

A type of oscillator due to Mr. W. W. Hanscom, operates with the gap surfaces immersed in alcohol. Their separation is automatically

FIGURE 59—One-half kilowatt Hanscom radiophone transmitter.

regulated by an electro-magnet plunger, a gravity adjustment by means of a sliding weight being provided for initial installation. The gap voltage is low (of the order of 100 volts). It is stated that steady automatic operation for hours has been secured. Only an occasional supply of alcohol and infrequent renewal of the gap surfaces are required.

In Figure 59 is shown such a set. The gap and regulator are mounted to the rear of the panel. The electromagnet winding is also used as a choke coil in the supply circuit. Direct current at voltages from 110 to 500 is supplied, and currents from 5 to 8 amperes pass

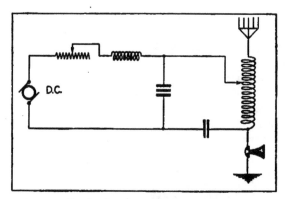

FIGURE 60—de Forest radiophone transmitter—
D. C. type.

through the gap. The system has been operated on wave-lengths between 300 and 2,700 meters. For modulation, a water-cooled microphone transmitter carrying 2.5 amperes is used.

With vacuum valve reception, distances of 100 miles (160 km.) are covered, but it is claimed that distances of 260 miles (400 km.) are occasionally bridged. On one occasion, the 800-mile (1,300 km.) span from San Francisco to Seattle was covered.

Dr. Lee de Forest has done considerable work in connection with radio telephony. Originally he worked with a small arc of the Poulsen type, and communication over short ranges was obtained. More recently

FIGURE 61—de Forest D. C. radiophone transmitter.

he has worked with several types of radio-frequent spark radiophone transmitters, and two of these types will be here described.

The first of these is a moderately high voltage, direct current system. The wiring diagram is given in Figure 60. As will be seen, a 1,000-volt, direct-current generator supplied a two-section quenching gap through a regulating resistance and choke coil. The gap itself is made of parallel studs of tungsten in air, with minute but regulable separation. Shunted around the gap is an oscillating circuit which is directly coupled to the antenna. Two heavy current microphones

FIGURE 62—de Forest portable radiophone set.

(sometimes air cooled by a blower) are connected in series in the ground lead of the antenna. A small set of this type is shown in Figure 61. It differs from that just described in that only one gap section is used and a single microphone in the antenna. The antenna ammeter is shown mounted on the upper left-hand portion of the apparatus box which contains the primary condenser, inductances, choke coils, and antenna switch. This sending-to-receiving transfer switch is controlled by the projecting knob on the upper right-hand portion of the apparatus box. The small 600-volt generator is shown separately. A 0.25-h.p. (200-watt) motor is recommended for driving the generator. The range is given as from 7 to 15 miles (10 to 25 km.). The set, as designed, operates at wave-length from 400 to 1,000 meters.

A portable type of radiophone is shown, set up, in Figure 62. It will be seen that the double microphone transmitter is used in the set in question. The receiving set is seen at the left and toward the back of the instrument case. A somewhat larger set is illustrated in Figure 63, with an air-cooled, two-section gap. The antenna switch and direct coupling coil are mounted to the right of the panel. When used for radio telephony, an air-cooled, twin-microphone transmitter is mounted on the panel, usually under the supply circuit ammeter.

FIGURE 63—de Forest 2 k.w. radiophone panel transmitter.

An alternating current system of spark radio telephony has been developed by de Forest. The circuit diagram is given in Figure 64. G is a 3,000-cycle alternator which supplies current to the primary of the transformer through the tuning condenser indicated, this latter having a value of approximately 8 microfarads. The transformer raises the terminal voltage from 100 to 5,000 volts. A number of gap sections similar to those previously described are used, and the primary is inductively coupled to the antenna. A double microphone is used in the ground lead as before. The audio frequency tuning to 3,000 cycles in the supply circuit is of interest. No data is available as to the extent to which the 3,000 cycle note can be eliminated and prevented from interfering with the speech in the arrangement under consideration. It is likely, however, that a square generator-wave form would be of assistance in this connection.

When it is attempted to receive signals from the de Forest radiophone transmitters by ordinary beat reception, (no speech being transmitted) a very poor note almost without musical characteristics is obtained. This is accounted for by the absence of a definite inverse charge frequency and the consequent extremely frequent alterations in phase of the radiated energy.

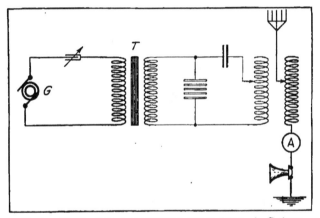

FIGURE 64—de Forest radiophone transmitter—A. C. type.

A 1-k.w., direct-current equipment placed on a train of the Delaware, Lackawanna, and Western Railroad permitted communication from Scranton to a moving express train at full speed up to 53 miles (85 km.). De Forest gives some interesting figures as to the average range of the sets. For the 2-k.w. set, using masts 100 feet (30 m.) high and at least 50 feet (15 m.) apart, the range over sea is up to 100 miles (160 km.) and over land up to 75 miles (120 km.). If 40-foot (13 m.) masts are used, these ranges are reduced to 0.3 or 0.4 of the values given. For the 5-k.w. sets, with similar 200-foot (60 m.) masts at least 100 feet (30 m.) apart, the sea range is up to 400 miles (640 km.) and the land range up to 300 miles (480 km.). This range is reduced to one-half the values given if the masts are reduced in height to 100 feet (30 m.). It is further stated that over heavily wooded and mountainous country, the ranges may be reduced 25 or even 50 per cent.

Excellent results have been obtained with a recent method of radio-frequent spark type using the Moretti "arc." The Moretti arc seems to be the most powerful generator of this sort yet discovered. It is a simple device, being shown in Figure 65. In the figure, the arc is shown enclosed in an air-tight box of insulating material, but this enclosure is not essential. The arc may be used in the open air. Both electrodes are of massive cop-

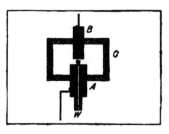

FIGURE 65—Diagrammatic representation of Moretti arc.

per, one with a plane surface and the other A with a longitudinal perforation through which is pumped a steady stream of acidulated water. This jet impinges on the upper electrode (which is the negative one, usually); and the velocity of the stream of water can be suitably regulated by a valve placed in the feed pipe. The theory of its action, as given by Professor Vanni, makes it a device somewhat analogous to the usual Wehnelt interrupter. He suggests that at the moment of formation of the arc, the water passes into the spheroidal state, vaporizing rapidly, and thus breaks the circuit very suddenly. At the same instant, the water is partly dissociated into hydrogen and oxygen; which, being an explosive mixture, quickly recombines, whereupon the entire cycle is repeated.

FIGURE 66—Scheidt-Boon Moretti arc as used at
Laeken station of Mr. Robert Goldschmidt.

Whatever the action, the effect is to open the arc circuit at a radio frequency, which fact can be verified by an examination of the arc by a rotating mirror oscillograph. The spark frequency is thus found to be several hundred thousand per second. As in the Chaffee arc, the impulses are stated to be unidirectional, though whether an inverse charge frequency exists and whether syntony to wave form is evidenced is not indicated in the published descriptions.

This arc has been improved in construction by Mr. Bethenod in that a precision regulator of the arc length has been designed by him, and that a special direct current generator has been used of high no-load e.m.f. and markedly lower load voltage. In this way, the series resistance in the supply circuit can be avoided and better efficiency attained.

FIGURE 67—Laeken (Brussels) station of Mr. Robert Goldschmidt, showing Moretti arc and Marzi microphone.

As normally used, the arc is placed in series with resistance and inductance across the terminals of a 600-volt direct current generator. The energy supply in the following experiments carried on by Professor Vanni was 1 kilowatt. Across the arc is placed a usual oscillatory circuit, which is inductively coupled to the antenna. In the antenna was placed Vanni's special hydraulic microphone transmitter to be described hereafter. Unquestionably, the remarkable results obtained are in large part to be ascribed to the development of this unusual form of telephone transmitter. The antenna current secured was 12 amperes.

In 1912, experiments were carried on by Vanni from the station at Cento Celle, several kilometers from Rome. The Island of Ponza, 120 km. (75 miles) away, was first reached, then Magdalena, 160 km. (100 miles) away; then Palermo, 420 km. (260 miles) away; then Vittoria, 600 km. (375 miles) away, and finally Tripoli, no less than 1,000 km. (625 miles) away. The results are noteworthy, and seem to be attainable without excessive uncertainty, as evidenced by the work done by Mr. Goldschmidt (of Laeken, near Brussels, in Belgium), and by the Marzi brothers in Italy.

The experiments carried on at Laeken early in 1914, before the unfortunate destruction of the station by its owners to prevent it from falling into the hands of an invading army, are of considerable interest.

As generator, a modified Moretti arc was used, fed with 600 volts. It is shown in Figure 66*. One electrode was rotated rapidly. This was the positive electrode and consisted of a number of discs mounted on an axle. The negative electrode consisted of the surface of rods held in sleeves with screw adjustment so that the arc length was directly regulable. As stated previously, a water jet was injected into the arc A special microphone heavy-current transmitter devised by the Marzi brothers was used, and this will be considered hereafter. Several Moretti arcs in series have been used by the Marzi brothers. With four arcs in series, running at 2,400 volts, radiophone transmission was effected between La Spezzia and Messina, the full length of Italy.

The equipment used in the Laeken experiments is shown in Figure 67. On the center of the table is mounted the Moretti arc, to the left of which are seen the coupling spirals. In the upper left-hand portion of the picture is shown the heavy-current transmitter, which is, in fact, controlled by the small transmitter held in the hand of the experimenter.

On March 13, 1914, using 3 amperes in the antenna, communication was established between the station at Laeken and the Eiffel Tower in Paris, a distance of 200 miles (320 km.). Tests were carried on regularly on wave-lengths of 300, 600, 800, and 1,100 meters. This arc shows the usual radio-frequent spark characteristic of satisfactory operation on short wave-lengths.

Reception was accomplished in various way, but it is interesting to note that the experimenters give the following as the order of merit of detectors in radiophone reception: sensitive crystals (such as galena), the audion, the Fleming valve, carborundum, and the electrolytic detector.

---

* Figure 66 and 67 are reproduced by permission from the French Journal "T. S. F.," based on material received from Mr. Scheidt-Boon of Brussels (1914).

It has been shown in connection with Figures 32 and 33, that a series of short, highly damped currents spaced regularly in a primary circuit would produce what was practically sustained alternating current of radio frequency in the secondary circuit (page 45). Suppose then that there be produced in a number of primary circuits in succession the currents shown in Figure 68 on the lines marked D_1, D_2, D_3, and D_4.

As will be seen, in the first primary circuit there are regularly spaced, highly damped wave trains. In the second primary, there are similar currents, but these occur at times which are later than the first primary trains by one-quarter of the time between trains. Similarly,

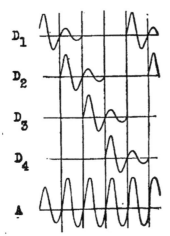

FIGURE 68—Production of Sustained Radiation by Marconi "Timed Spark" Method.

in the third primary, there are wave trains which lag behind those of the first primary by a time equal to two-quarters of the time between trains; and in the fourth primary, the trains lag by three-quarters of the time between trains behind those of the first primary. The current which is produced in the common secondary, to which all the primaries are coupled is shown in the line marked A in the figure. It is clear that after the first few periods, the secondary current will have practically constant amplitude. It will be noticed also that for perfect regularity in secondary current, the time between wave trains should be integrally related to the radio frequency period of the secondary current.

To carry out this idea, Senatore Marconi has devised the circuits shown in Figure 69. The first primary circuit consists of the rotary discharger D_1, the condenser C_1, and the inductance P_1. The second primary circuit consists of the discharger D_2, and P_2 and C_2; and so on.

Each of the dischargers is fed with *direct* current of high voltage (e.g., at 10,000 volts) by the generator G. The antenna circuit (between A and G) is coupled to the various primaries through the inductances S_1, S_2, S_3, and S_4.

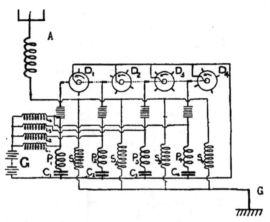

FIGURE 69—Marconi "Timed Spark" Method of Producing Sustained Radiation.

By the means shown, practically sustained radiation is obtained in the antenna circuit. It will be noted that each of the dischargers is shifted on the common shaft by an angular distance equal to one-quarter the angular distance between successive studs on the dischargers.

A later modification of the method given is used at the Carnarvon, Wales high power station of the Marconi Company, which station will work with a corresponding station at Marion, Massachusetts. It is

FIGURE 70—Marconi Company "Timed Spark" Discharger.

noticeable in receiving Carnarvon in the United States, by beat reception, that pure beat tones are obtained; thus demonstrating the essentially constant amplitude and frequency of the radiated waves. The multiple discharger used in this way at Carnarvon is illustrated in Figure 70. The driving motor is visible at the right, and the separate "trigger" dischargers at the left. Though equipment of this sort has not been used for radio telephony up to the present, so far as the Author is aware, it does constitute an apparently reliable and powerful method of producing the requisite sustained radiation and may be applied in the field mentioned in the future.

CHAPTER IV.

(c) Vacuum Tube Oscillators; Dushman's Data; Temperature and Space Charge Limitation of Plate Current; Thermionic Currents in Filament; White Filament Supply Method; Grid Potential Control of Plate Current; Tube Amplification of Alternating Current; Self-Excited Oscillations; Oscillating Circuit of Meissner; Marconi-Franklin Circuits; de Forest Ultra-udion and Other Circuits; High Power Tubes; General Electric Company Pliotrons; Oscillating Circuit of General Electric Company; Western Electric Company Tubes; Experiments of Colpitts; Experiments of Heising; Nicolson Tube; General Electric Company Dynatron and Pliodynatron; Hull's Dynatron Amplifier and Oscillator; Hull's Pliodynatron Controlled Oscillator.

(c) VACUUM TUBE OSCILLATORS.

There has arisen within the last few years a new and important type of sustained radio frequency generator, namely, the hot cathode vacuum rectifier, usually with three internal electrodes. As will appear, the ease and certainty of control of currents formed by pure electron streams in a vacuum has rendered these devices suitable not only for use as generators, but also amenable to telephonic modulation and control of the radio frequency output. In the following discussion, however, we shall consider only tube construction and the associated circuits enabling the generation of radio frequency currents. The modulating methods for radio telephonic purposes will be considered together with the station apparatus under a later heading.

Since the mode of action of the devices described here is still, in many cases, under judicial consideration in the courts of this country, we shall confine ourselves to giving without comment the explanations advanced by the various investigators.

We shall consider first electron currents through a vacuum. If the filament FF in Figure 71 is heated to bright incandescence by the filament battery FB (regulated, if necessary, by a series rheostat in the battery circuit, not shown) there will be emitted from the filament a copious stream of negative electrons that is, small charges of negative electricity. A definite number of these are emitted from the filament per second for each centimeter of length of the filament. The number emitted depends markedly on the temperature, increasing extremely rapidly as the higher temperatures are attained. For example, Dr. Saul Dushman of the General Electric Company, found that the current per square centimeter of filament surface increased from about 0.14 ampere per sq. cm. at 2,300° absolute to 0.36 ampere per sq. cm. at 2,400°. The values for 2,500° and 2,600° were respectively 0.89 and 2.04 amperes per sq. cm. It is quite obvious that the highest temperatures of filament consistent with not burning out the filament and a reasonably long filament life are desirable if large currents are to be passed through the tube.

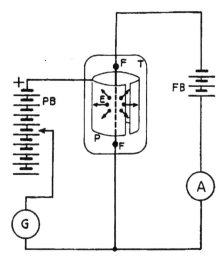

FIGURE 71—Thermionic currents.

Suppose that the cylindrical metal plate be placed around the filament as indicated at P. Suppose further that a battery, PB, and galvanometer G be connected in series between plate and filament. If the negative side of the battery be connected to the plate, practically no current will flow through the galvanometer. If, on the other hand, the positive side of the battery be connected to the plate, negative electrons will be attracted to the plate, returning to the filament at the lower point, F. Using the ordinary convention for the direction of current flow (which is opposite to the direction of flow of the electron stream), we say that a current flows from the plate to the filament. The device is therefore a rectifier, since it permits the flow of current from plate to filament, but not *vice versa*. This form of the device has been used by Fleming since 1906 as a detector for radio receivers. In a highly evacuated form, it has recently been developed into the new Coolidge X-ray tube and the so-called "kenotron" or high voltage, high vacuum rectifier of the General Electric Company.

The current through such a device in the plate circuit obviously depends on the plate potential. In general, the more positive the plate, the higher the electron velocity across the space between filament and plate, and the greater the plate current. There is however, a clear limitation to this increase of current. At any given temperature, only

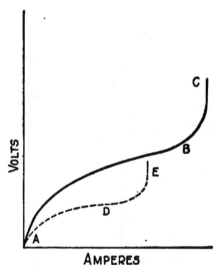

AMPERES

FIGURE 72—Relation between voltage and current for pure electron rectifier at a given temperature.

a given number of electrons can be emitted by the filament per second, and when all of these are drawn to the plate per second, no increase in plate voltage will cause an increase in plate current. This is called the *temperature limitation* of plate current. In Figure 72, it is illustrated at *B*. In the lower portion of the curve the current increases (as can be shown by mathematical analysis) with the three-halves power of the applied plate voltage, but at *B* we reach the limiting current value at the given temperature and the curve bends sharply to *C*, whereafter the plate current remains constant unless the temperature of

the filament is raised. In the portion AB of the curve, the current from
the plate to filament is actually given by the equation:

$$i = 14.65(10)^{-6}\frac{l}{r}e^{3/2}$$

where i is the current in amperes in the plate circuit, l is the length
of the filament in centimeters, and r the radius of the cylinder in centimeters. The curve ADE is for a lower temperature, and therefore also
for a lower limiting current.

There is a second type of current limitation at a given plate
voltage which may prove very serious in practice in high vacuum
tubes. This is the so-called *space charge limitation,* and depends on

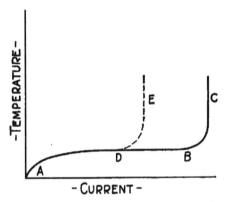

FIGURE 73—Space charge limitation of
thermionic current at a given
plate voltage.

the following considerations. If the plate voltage has a given value,
increase of filament temperature will increase the plate current to a
point B, but not further. This is due to the following effect: The
cloud of negative electrons surrounding the filament at any time acts
as a large negative charge in its neighborhood, and consequently
repels all electrons which are, or tend to be, emitted by the filament,
thus choking back the electron current stream. If the charge in the
space surrounding the filament becomes sufficiently great, no increase
in temperature at a given voltage will produce any further current.
Either the plate voltage must be increased or the bulb construction
altered so as to diminish the space charge. Bringing the plate and
filament close to each other will diminish the space charge effect. The
effect is indicated at B in Figure 73; and, for a lower applied plate
voltage, at D with the dashed line.

In considering the current-carrying capacity of vacuum tube rectifiers, Dr. Dushman gives data as to the current in milliamperes per centimeter of filament length at a safe working filament temperature. Thus with a filament 0.005 inch (0.012 cm.) in diameter, 0.030 ampere can be safely emitted per centimeter of length. Under such conditions, the filament heating current will represent 3.1 watts of power per centimeter of length. For a filament 0.01 inch (0.025 cm.) in diameter, these figures become respectively 0.10 ampere and 7.2 watts per unit length. This gives an indication of what may be expected from tubes of ordinary dimensions based on these thermionic currents.

A curious effect is encountered when the joint filament heating and thermionic (pure electron) currents are combined. In the filament heating circuit shown in Figure 74, the current circulates in the direction indicated by the dotted arrows. Under normal conditions, therefore, the ammeters A_1 and A_2 read the same. If, however,

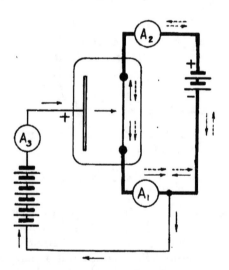

FIGURE 74—Illustrating Combined Lighting
and Thermionic Currents.

the plate circuit is closed, and a current indicated by A_3 appears in that circuit, its direction of flow will be as indicated by the full line arrows. (It is understood that the direction of current flow is opposite to that of the negative electrons, in accordance with the commonly accepted convention). It will be noticed that the plate current A_3, will flow *outward* from *both* ends of the filament. Consequently, at the lower end it will assist the lighting current, while at the upper end it will oppose it. So that, if A is the true lighting current, the readings of the ammeters will be given by $A_1 = A + A_3$ and $A_2 = A - A_3$. With small tubes, such as might be used for receiving, this effect is of no practical importance, but on larger, heavy plate current tubes (with filaments already worked near the burn-out point) it may become serious.

This effect has been ingeniously minimized by Mr. William C. White, to whom much of the recent development of the pliotron is due, through the use of the circuit shown in Figure 75. Here the filament is lit by the alternating current from the secondary of the transformer T. The connection of the plate circuit is made to the middle of the supply secondary winding. A similar method might be applied to connection to the middle

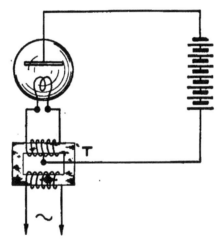

FIGURE 75—General Electric Company-White method of plate circuit connection.

point of a storage battery (or three-wire direct-current generator) used for the supply of lighting current.

We have assumed so far that the vacuum within the bulb was practically "perfect"; that is, a few ten-millionths of a millimeter of mercury or less. Furthermore, by the use of elaborate exhausting and internal heating methods, it is assumed that the electrodes have been thoroughly freed from any occluded gases so that the tube will remain constant in operation. (See Dr. Langmuir's paper appearing in the September. 1915. issue of the "Proceedings of the Institute of Radio Engineers.") Such perfection of vacuum is not easily obtained or maintained, and tubes containing or evolving gas will show markedly different effects from

those described. In the first place, the current between plate and filament will be much increased. The reason for this is the following:

The rapidly moving electron stream will ionize by impact the gas
molecules; that is, dissociate the atoms into negative electrons *and positive
ions*. These positive ions will recombine with the "electron cloud" surrounding the cathode, thus neutralizing and destroying the effects of the
space charge. In consequence, tubes in which gas (and consequently
positive ions) are present will pass greater currents at low plate voltages
than will the extremely high vacuum tubes. Among tubes having present
positive ions (and diminished space charge effect) are the original de
Forest audions and the von Lieben-Reisz oxid filament tubes. At first
sight, it might seem that the presence of positive ions and increased current in the plate circuit was an unmixed advantage, and there is no
doubt that it constitutes a convenience in ordinary detector tubes in
that it permits the use of comparatively low plate voltages. On the
other hand, it has at least two marked disadvantages.

The first of these is the fairly rapid filament deterioration of such
tubes when any considerable plate current passes. The presence of
positive ions leads to ionic bombardment of the negatively charged filament. The positive ions are comparatively massive (in relation to the
negative electrons); and when they strike the filament at fairly high
velocities, the surface is rapidly damaged. This is not at all the case for
the high vacuum "pure electron discharge" tubes, where positive ions
are not present. Furthermore, when used to pass any considerable
amount of plate current, the gas-containing tubes may become dangerous
in that the gaseous ionization may rise to the familiar "blue glow" point.
At this point continuous and progressive ionization of the gas occurs
together with greatly increased plate current. While they may not be
much more than an inconvenience with small tubes, with large tubes at
high plate voltages it may lead to disastrous currents and consequent
violent tube destruction. For these reasons, very high vacua are generally
desirable in tubes.

It is a fact, though not well known, that the usual Fleming valve or
rectifier can be used to produce sustained oscillations when shunted by a
circuit of large inductance and small capacity without any third electrode
or control member. This method is not used in practice because of the
high voltages required, the troublesome large resistances in the feeding
circuit, and the very rapid deterioration of the tube and its irregular
operation.

For the production of sustained radio frequency oscillations from
vacuum tubes, a third or control member may be employed. This may
be in the form of a perforated plate or a grid of wire placed between the

plate and filament so that the electron stream must pass through the meshes of the grid. The remarkable mobility of the electron stream permits of ready control of the current between plate and filament. Dr. Langmuir has stated that the current between plate and filament with the control member inserted is given by the equation:

$$ i = 14.65(10)^{-6}\frac{l}{r}(e + k\,e')^{3/2} $$

where i is the current in amperes in the plate circuit, l the length of the filament in centimeters, r the radius of the surrounding plate (of cylindrical form) in centimeters, e the voltage in the plate circuit, e' the grid potential (relative to the filament), and k a constant. The constant, k, is dependent on the spacing of the grid wires, the distance of the grid from the plate and filament and the construction of the tube. Roughly speaking, the finer the spacing of the grid wires, the larger the constant k and the smaller the grid potential variations which will completely control the plate current. The danger with fine grids is that small positive potentials will produce excessively large plate currents. With a coarse grid, the control voltages must be larger, but the danger mentioned above is minimized.

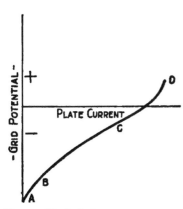

FIGURE 76—Relation between grid potential and plate current for pure electron amplifiers.

The control energy required for producing the requisite grid potential variations is quite small and herein lies the remarkable amplifying (and oscillating) power of the device. Aside from grid leakage and grid charging currents there are no sources of energy loss in the grid circuit inside the bulb.

A typical grid potential-plate current curve is given in Figure 76. It will be seen that for large negative grid potentials (at A) practically no current flows in the plate circuit. From B to C the current through the plate circuit varies practically linearly with the applied grid (negative) potential, and it is in this range that the tube should be worked for radio telephonic oscillation or control. At C, the plate characteristic begins to flatten, until at D practically no further increase of plate current can be produced by more positive grid potential. The flattening of the curve at D may be caused either by temperature or space charge limitation of the plate current and determines the rating of the tube at a given plate voltage.

In Figure 77 is illustrated the mode of action of the electron relay as an amplifier of alternating current. The alternator, A (which may, of course, be replaced by the oscillating circuit condenser terminals), is connected to the grid and filament of the tube. The plate circuit is supplied by the battery B which, we shall assume, readily permits the passage through it of alternating current. If this last is not true, a large condenser must be shunted across the battery, thus by-passing the alternating current without interfering with the direct plate current. In series with B are connected the direct current ammeter A_1, the alternating current ammeter, A_2, and the primary of the transformer, T. It

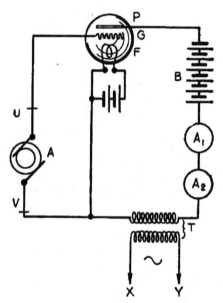

FIGURE 77—Amplification of alternating
current energy.

is assumed that A_1 does not impede the flow of alternating current in the plate circuit; otherwise it may have a condenser placed in parallel with it. The secondary terminals, X, Y, of the transformer T constitute the output terminals of the amplifier or "repeater."

Under the conditions shown, the plate current will remain at the steady value indicated by AB in Figure 78 so long as the alternator, A, is not running. The effect of closing the alternator circuit is shown at BC in Figure 78. In the figure the median value of the portion, BC, is taken as equal to that of AB; that is, it is assumed that the fluctuating current swings up and down around an average value equal to the original direct current. This is generally not the case; since grid circuit rectification, flattening of the grid potential-plate current characteristic, or occasional positive grid charges may cause the average plate current to go up, remain fixed, or drop when the alternating potential difference is applied to the grid and filament. In

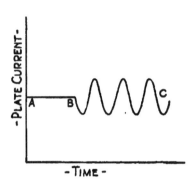

any case, however, the pulsations in current in the plate circuit will be marked if the grid potential variations are sufficient, and there will be available at the terminals, *X*, *Y*, the amplified energy. As shown, the device may obviously be used as an audio or radio frequency amplifier, and is indeed so employed respectively in the trans-continental wire telephone lines and in ordinary receiving radio sets.

FIGURE 78—Plate current-time curve.

It has been pointed out that the energy delivered at the terminals, *X*, *Y*, is many times greater than that required at the terminals, *U*, *V*, of the alternator. For example, there may be available at *X*, *Y*, 10 watts, while only 1 watt is required at *U*, *V*. It would immediately seem that if one of the 10 watts available at *X*, *Y*, were transferred back to *U*, *V*, by

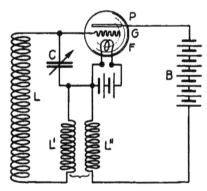

FIGURE 79—Oscillating circuit.

coupling or otherwise, the alternator might be removed, but the system would continue to sing or oscillate steadily as a generator of alternating current. A typical circuit arrangement, shown by E. H. Armstrong, for

securing this so-called "regenerative coupling" is given in Figure 79.*
It will be seen that the arrangement is similar in principle to Figure 77,
except that the alternator, *A*, has been replaced by the oscillating circuit,
L L' C, or rather by the condenser terminals of *C*. In addition, there

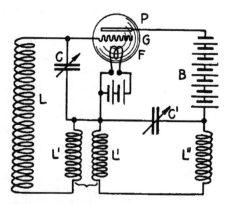

FIGURE 80—Plate circuit tuning in oscillating circuit.

has been added the coupling, *L' L"*, between the grid circuit, *L L' C*, and
the plate circuit, *L" B*. A system such as that shown will oscillate vigor-
ously if the circuit constants are properly chosen. The output energy is
in general obtained by coupling to a coil inserted in the plate circuit. It
is this type of oscillator, which, used as a detector also, is so directly applicable to long distance beat reception; and has accordingly been widely
applied for that purpose.

An improvement on the simple circuit of Figure 79 has been shown
by Armstrong, and is given in Figure 80.* It contains an added inductance, *L"*, in the plate circuit and a condenser, *C'*, across the terminals
of *L'* and *L"* whereby the plate circuit may be tuned to the same frequency as the grid circuit or approximately so. The efficiency and output
of the oscillator are generally increased by such an arrangement; but,
on the other hand, the complexity of apparatus and difficulty of adjustment may sometimes become undesirable.

In working with the various types of oscillating circuits to be shown,
it is quite essential that the grid connection shall be to such a point of
the conjoint grid and plate circuits that the electromotive forces placed
on the grid are in the proper phase relation to the alternating current

---

* "Proceedings of the Institute of Radio Engineers," Volume 3, number 3, September,
1915.

produced in the plate circuit, otherwise the system will not persist in oscillation.

A form of oscillating circuit of simple electrical nature, due to Dr. A.

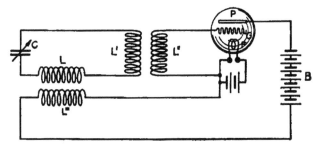

FIGURE 81—Meissner oscillating circuit, 1913.

Meissner of the Telefunken Company, and invented before March, 1913, will be next considered. The circuit is shown in Figure 81. It will be seen that the grid and plate circuits are coupled, but indirectly, through

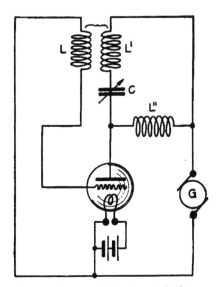

FIGURE 82—Arco-Meissner oscillating circuit.

the tuned circuit L L' C. The inductance, L, of this circuit is coupled to the plate circuit, while the inductance, L', of the same circuit is coupled to the grid circuit. In consequence, sustained alternating current will be produced in the circuit, $L\ L'\ C$, as previously indicated.   In practice, resistance may be inserted in the circuit, $L\ L'\ C$, for absorbing the output of the system; and in fact, the capacity, C (and the resistance just referred to), are replaced by the antenna when radiation is desired.   Another form of circuit used by the same company, and the joint invention of Count Arco and Dr. Meissner in 1914 is shown in Figure 82.   It

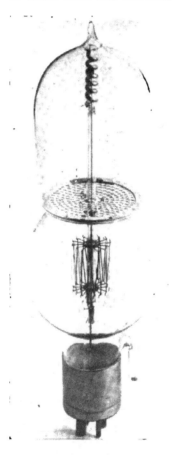

differs from that previously shown in that the intermediate coupling circuit is replaced by a direct inductive coupling between grid and plate circuits.   This coupling, $L\ L'$, links the grid circuit to the tuned, absorbing plate circuit, $L'\ L''\ C$, which, as before, may either contain the antenna or be coupled thereto.

An interesting type of bulb was used by Dr. Meissner in his experiments; and a photograph of this bulb is shown in Fig. 83.   Bulbs of this sort give current amplifications up to thirty times.   It must be at once mentioned that these are *not* high vacuum bulbs, an atmosphere of mercury vapor being purposely provided by the small piece of mercury amalgam shown sealed into the small side tube at the bottom of the tube.   The result of this vapor and the oxide-coated Wehnelt (heated) cathode is that the tube in operation shows a continuous blue glow.

As has been stated, the filament is a platinum strip, about a meter (3 feet) long in all, 1 mm. (0.04 inch) wide, and 0.02 mm. (0.002 inch) thick.   It is thinly coated with a mixture of calcium and barium oxides, and is brought to a bright red heat

FIGURE 83—Lieben tube of
Telefunken Company.

by a current of about 2 amperes from a 28 to 32 volt storage battery, the current being regulated by a 5 ohm variable series resistance.   Considerable heating power is, therefore, required; and the source of this power must be an extremely constant one.

The plate circuit is fed from a 220-volt source which may be an ordinary dynamo with choke coils in the supply leads to cut down the incidental noises. The plate circuit current is about 0.01 ampere, and the dark space interrupting the blue glow above the grid can be used for rough indication of the current through the plate circuit. As will be seen, the plate itself is of heavy aluminum wire.

The grid is a perforated aluminum sheet, the size of the perforations being about 3.5 mm. (0.14 inch). It will be noted that all connections to this bulb are made through the bayonet socket in the base, this being so arranged that the bulb can be placed in its socket only in the correct position. The life of these tubes is claimed to be 1,000 hours or more.

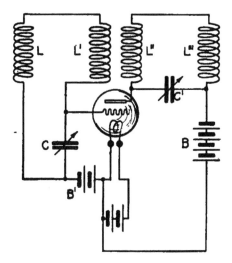

FIGURE 84—Marconi Company-Franklin
Circuit, 1914.

When used as an oscillator, wave lengths as short as five or ten meters have been obtained, and with great constancy. Using a plate voltage of 440 (instead of the usual 220), twelve watts have been transferred to an antenna, corresponding to an antenna current of 1.3 ampere in a 7-ohm antenna at 600 meters wave-length.

One of the circuits devised in 1914 by Mr. Franklin of Marconi's Wireless Telegraph Company of England is shown in Figure 84.* It will be noticed that the plate oscillating circuit is tuned by means of the condenser, C', and that one of its inductances, L'', is coupled to the grid circuit inductance, L'. The grid circuit, L L' C, is also tuned. Energetic

* British patent, No. 13,248, of 1914.

oscillations can thus be obtained. It will be noticed further that there is included in the circuit between filament and grid the battery, B'. The purpose of this battery is to enable choosing such normal grid potential as shall give a desired plate current through the bulb, and desired output with high efficiency. Indeed, it is necessary with most bulbs to keep the grid at a negative potential, since, if the grid becomes positive, current begins to flow from the grid to the filament with consequent absorption of energy in the grid circuit. The amplifying action of the tube and its efficiency as a sustained current generator are then impaired.

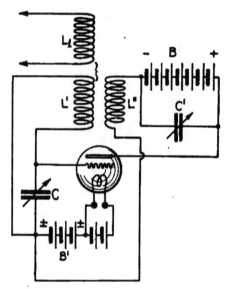

FIGURE 85—English Marconi Company oscillating circuit; modified form.

In Figure 85 is shown a simplified diagram of another form of transmitting circuit used by the English Marconi Company in its ship radiophone transmitters. The details of the wiring diagram will be given under "Control Systems." It need only be mentioned that the alternating current energy is withdrawn from the oscillator at L_1.

Dr. de Forest has carried on extensive experiments with vacuum tube oscillators. One of the earliest and simplest circuits is his "ultraudion" circuit, shown in Figure 86. It is normally used in receiving, though it is naturally available also for generation of greater power. As shown, the telephone T and battery B in the plate circuit are shunted by a "bridging condenser" C''. Connected between the plate and grid is the oscillating circuit, *L C*, one side of which is directly connected to the plate, and the other to the grid by the small condenser, *C'*. This condenser is usually shunted by a leakage resistance (not shown in the figure)

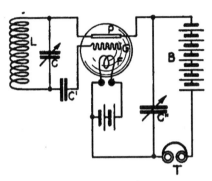

FIGURE 86—de Forest ultraudion circuit.

which prevents the accumulation of an excessive negative charge on the grid and consequent limitation of the plate current.

Dr. de Forest explains the action as follows: ''There is only one oscillating circuit. This circuit is such that a sudden change of potential

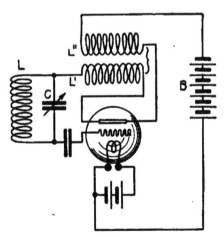

FIGURE 87—de Forest oscillating circuit, 1915.

impressed on the plate produces in turn a change in the potential impressed on the grid of such a character as to produce, in its turn, an opposite change of value of potential on the plate, etc. Thus the to-and-fro action is reciprocal and self-sustaining.'' In thus explaining the action of the device, Dr. de Forest takes sharp issue with Mr. Armstrong, who claims that the circuit is ''regenerative'' in the sense that there is an inductive-capacitive coupling between the plate and grid circuits, which latter circuits are claimed by Mr. Armstrong to be existent and clearly defined.

A later oscillating circuit (1915), due to de Forest, is shown in Figure 87. It will be seen that this circuit differs from the normal ultraudion in that there is a coupling added between the grid and plate circuits. This coupling is L' and L'' and is presumably intended to reinforce the production of oscillations and produced greater outputs in consequence. The coil, L'', is referred to as a ''tickler'' coil.

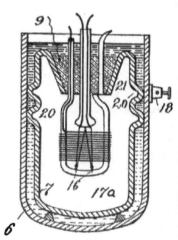

The question of considerable outputs from vacuum tube oscillators has led to the consideration of methods of heat-resistant tube construction. An attempt in this direction is shown in Figure 88 and is due to de Forest. The two metal vessels, 6 and 7, are so arranged that the space between them is filled by a heat-conducting fluid, e. g., mercury or certain oils. This fluid acts at the same time as a means of sealing the inner vessel and of preventing air leakage. The grid, filament, and plate structure are mounted inside the inner vessel in the usual manner. The inner vessel is corrugated in the region, 20, so as to provide plenty of heat conducting surface where this is most needed.

FIGURE 88—de Forest high-power tube construction.

The General Electric Company has developed a number of types of extremely high vacuum tubes and the circuits necessary for their use. One of the simplest of these, and one having marked advantages, is shown in Figure 89. Here both plate circuit L'' C'' G and grid circuit L' C' are tuned and coupled to each other. The output circuit is connected to the inductance, L, coupled as shown. A unique feature of the circuit is that the same generator, G, is used both for lighting the filament through the auxiliary regulating resistance, R, and for supplying the plate circuit directly. It is thus possible to connect such an arrangement directly to a single source of direct current and to start the oscillation by merely closing a single switch. Such automatic action is a desideratum in radiophone equipment.

The actual appearance of the General Electric pliotron or three-electrode tube is indicated in Figures 90, 91, and 92. The first of these figures shows the mode of mounting the filament and grid member of a pliotron. The "W" filament is suitably anchored and supported. The grid itself is wound on a tungsten frame. Figure 91 represents a later type of filament and grid support. This type has increased rigidity and is more heat resistant. In addition, the insulation has been improved, particularly with a view to resisting the extremely high temperatures attained within the bulbs when in operation. The appearance of one of the complete bulbs is clearly shown in Figure 92. The massive tungsten

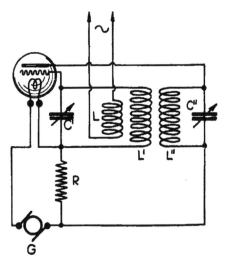

FIGURE 89—General Electric Company oscillating circuit.

plates are seen to be properly supported outside the filament grid structure, and from the opposite end of the tube. Tubes of this sort can stand thousands, and even tens of thousands, of volts between plate and filament without showing any blue glow due to gas present in the tube. The output of even a comparatively small tube of the type shown in Figure 92 runs into hundreds of watts at plate voltages of about one thousand volts. Such tubes and the circuits associated with them will be further considered under a later heading, wherein complete radiophone sets of the General Electric Company are shown.

We consider next certain phases of the work of the Western Electric Company. A circuit used for the production of oscillations by that company and due to Mr. Edwin Colpitts in 1915 is shown in Figure 93. The plate circuit is fed from the battery, B, which is in series with the choke coil or inductance, L_1. Consequently the plate voltage does not remain constant. The tuned plate oscillating circuit is $L'\ C'$, this being inductively coupled to the tuned grid circuit, $L\ C$. The grid is maintained at a negative potential by means of the battery, B', the oscillations impressed on the grid being prevented from passing through the battery, B', by means of the inductance, L_2. The output of the bulb is drawn from the coil, L'', which is inductively coupled to the inductance in the plate circuit.

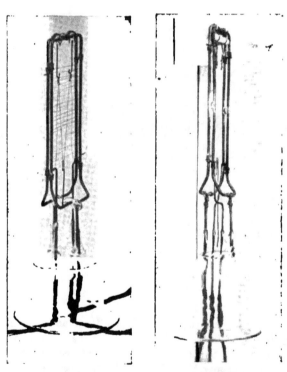

FIGURE 90—Filament and   FIGURE 91—Filament and
grid element of pliotron.   grid element of pliotron.

A line of development which the Western Electric Company, among others, has pursued in connection with the obtaining of considerable outputs has been the amplification of the output of a single oscillator by a bank or banks of vacuum tube amplifiers, these individual amplifiers being placed in groups in parallel. While apparatus of this type tends

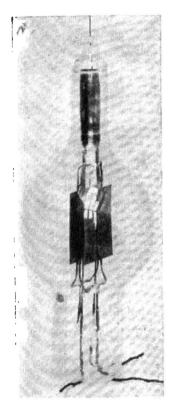

FIGURE 92—General Electric
Company pliotron.

to become bulky and clumsy when a very considerable number of bulbs are used, it has considerable electrical flexibility. An arrangement of this sort due to Mr. R. Heising is shown in Figure 94. Herein the oscillator A, is coupled inductively to the combined grid circuit of a number of amplifying bulbs, A', A'. The grids of these bulbs are maintained at a suitable negative potential by the battery, B'. The circuit, $L\,C$, is tuned to the oscillator frequency. The resistance, R (which is non-inductive), is shunted across C so that the sharpness of resonance of the combined circuit is adjustable and that its impedance at a definite frequency shall have a sharply defined value. As will be seen, all the grids of the amplifiers, A', are connected in parallel, as are also their plates. A common plate battery, B, feeds all of them. In series therewith is an inductance which is coupled to the circuit, $L'\,R'\,C'$, this latter being the input circuit of the second bank of amplifiers, A'', A''. In this way the amplified voltages which are produced in the plate circuit of the first bank of amplifiers are brought to the grids of the

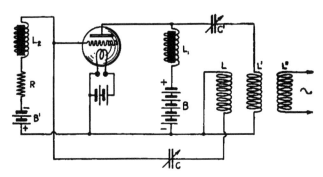

FIGURE 93—Western Electric Company—Colpitts oscillating circuit.

second bank of amplifiers. This second bank of amplifiers is intended for increasing the alternating current in the output circuit, whereas the first bank was intended primarily for a voltage increase. The resistance, R', is inserted in the grid input circuit of the second bank of amplifiers to render the operation more stable. The plate circuit of all the amplifiers, A'', are fed from the common battery, B'', and an inductance in this plate circuit is coupled to the antenna tuning coil, L''. By this means

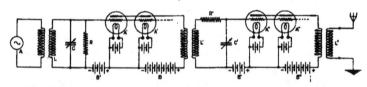

Figure 94—Western Electric Company-Heising oscillator—amplifier arrangement.

the amplified currents are set up in the antenna or final output circuit. This system will be further considered under another heading in connection with the radiophone work of the Western Electric Company.

The construction of vacuum bulbs for large outputs has engaged the attention of the engineers of this company as well. A well-defined trend of their development has been the attempt to secure very effective control

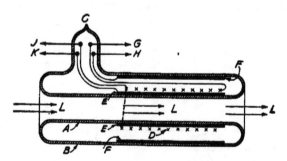

Figure 95 —Western Electric Company—Nicolson
high-power bulb.

by placing the filament and grid very close together. In fact, actual contact (though with an insulator, such as nickelous oxide, between) has been considered. The arrangements developed for this purpose will be considered in greater detail in connection with receiving apparatus. For transmitting work Mr. A. Nicolson has developed the type of bulb shown in Figure 95. A glass tube, A, of cylindrical form, is sealed inside another cylindrical glass tube, *B,* and the space between is exhausted through the seal *C.*

Prior to the exhaustion, the filament, grid, and plate members are inserted or slid into the space between the inner and outer tubes. The filament is a twisted platinum strip coated with nickelous oxide and wound around the metal cylinder *E,* which is the grid. The filament, *D,* is represented by the two lines of crosses along the length of the cylindrical grid. The filament terminals are brought out of the tube through the leads, *J* and *G.* It will be noticed that the grid is *internal* to the filament in this particular tube, a comparatively rare construction. The grid lead out of the tube is *K.* The plate is the outer cylinder, *F,* and its connection to the outside of the tube is *H.* The plate is inserted into the tube at the same time as the grid and filament, that is, before exhaustion.

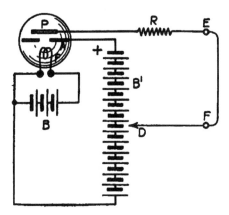

FIGURE 96—General Electric Company—Hull dynatron amplifier circuit.

Cooling of the tube is accomplished by passing a liquid or gas through the central orifice as indicated by the arrows, *L.* The exterior portions of the tube are similarly cooled, and this is claimed to enable the tube to operate continuously with heavy plate currents.

Through the courtesy of the General Electric Company and Dr. Albert W. Hull, we are enabled to present to our readers a more recent development in vacuum tube amplifiers and oscillators, namely "the dynatron." This device depends on a principle hitherto not used in this connection, namely secondary emission. This phenomenon is as follows: When a stream of rapidly moving negative electrons falls on a metal plate, if the velocity of the stream is not very great, no unusual effect will be noticed. If the velocity is somewhat increased each electron impinging on the plate will liberate from the molecule which it strikes one slowly moving electron. As the velocity of the impinging or "primary" electron stream is increased, at each collision, two electrons will be liberated from the plate, and the number of "secondary" electrons liberated by each primary electron on impact may be as many as twenty for very high primary electron velocities.

Let us now consider the arrangement of circuits shown in Figure 96. The bulb shown is a dynatron, containing an incandescent filament, F, an *anode*, A (which is a perforated plate), and the plate, P. It is at once to be noted that the anode, A, is *not a grid*, being maintained at a *fixed* and high *positive* potential and not serving as an *input* member of the system. Unless this is kept in mind, the action of the device can not be

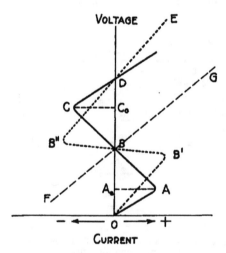

FIGURE 97—Dynatron voltage amplifier characteristics.

understood. The filament is maintained incandescent by the battery, B. Between the filament and the anode, A, is connected the battery, B', with its positive end connected to A. So far the device will act just as does an ordinary hot cathode rectifier, *e. g.*, a kenotron, with the exception that a great number of electrons moving from the filament, F, to the anode, A, will pass through the hole or holes in the anode and strike the plate. So long as the velocity of the electrons striking the plate, P, is not high, the curve connecting applied voltage (between the plate, P, and the filament, F) and the current flowing in the plate circuit (*e. g.*, between points E and F) will be similar to that for a kenotron. Suppose for the present the resistance, R, in the plate circuit to be zero. As long, then, as the tap, *D*, is so placed that the plate is not very positive, we get the usual characteristic indicated by the portion *OA* of the curve of Figure 97, which, as stated, resembles the normal current-voltage curve of a kenotron rectifier.

As we approach the point, *A*, of the curve, however (by raising the voltage of the plate by moving the tap, *D*, up the battery, *B'*), the electrons striking the plate begin to have higher velocities and secondary emission occurs. In consequence the electrons released by the secondary emission are produced in increasing quantities. Since the anode is *more positive* than the plate, these electrons will be attracted to the anode and there absorbed. As for the plate, it begins to lose by secondary emission an appreciable portion of the current which strikes it so that the net current in the plate circuit (*DFERP*) becomes smaller and smaller as the plate voltage is increased. This is shown in the portion, *AB*, of the curve of Figure 97, which shows that the current in the plate circuit is *diminishing* for *increasing* plate voltage. At *B* the plate loses as many electrons as strike it, and the net current is zero. From *B* to *C*, as the voltage of the plate is further increased, each electron that strikes the plate liberates more than one electron so that the plate on the whole *loses* electrons and the plate current is actually reversed and negative. At *C* the limit of re-emission is reached, and thereafter the plate current rises along the curve, *CDE*, as the voltage is increased. We have, in the range, *ABC*, of the applied plate voltage a most curious effect, namely that an increase of voltage causes an increase of current *in the wrong direction.* That is, between voltages A_0 and C_0 the plate-to-filament circuit of the dynatron acts as a true "negative resistance" which, so far from opposing the flow of current, actually assists it. It acts, therefore, in a manner very roughly analogous to the electrical (though not to the physical) behavior of the Poulsen arc and is capable of being an amplifier or oscillator. The arc, however, has a negative resistance characteristic only for *increasing* current, but acts as an open circuit for *decreasing* current. The dynatron has a stable negative resistance in either case. Furthermore, the dynatron has no hysteresis or lag, but responds instantaneously, because it does not depend on gas ionization, as does the arc.

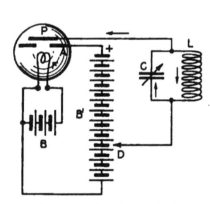

FIGURE 98—Dynatron oscillator.

To make the device a strong amplifier, we insert a resistance, R (in Figure 96), in the plate circuit, which resistance has a positive value nearly equal to the negative resistance of the dynatron plate circuit. The current-voltage curve of such a resistance will be parallel to the line FG in Figure 97, where FG slopes to the right nearly as much as AC to the left. The plate circuit characteristic of the dynatron will then become the curve $OB'B''E$, which is dotted in the figure. It will be seen that a very small change of voltage in the neighborhood of the value, B, will cause a very great change in the current in the circuit from B' to B''. The small exciting voltage would be inserted into the plate circuit, for example between the points E and F.

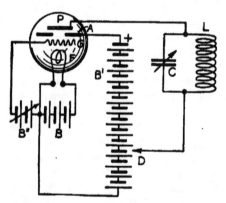

FIGURE 99—General Electric Company-
Hull pliodynatron controlled oscillator.

Since the dynatron is a negative resistance, it is essentially an unstable device and will, if an oscillating circuit is included in the plate circuit, produce in that oscillating circuit sustained alternating currents. The circuit diagram therefor is shown in Figure 98, which is quite similar to Figure 96, except that the oscillating circuit, LC, is added in the plate circuit. The directions of current while the capacity, C, is discharging are shown by the small arrows, and it will be seen that the capacity discharges partly through the inductance, L, and partly through the plate circuit of the bulb.

Using the dynatron as an amplifier, voltage amplifications of as much as 1,000-fold have been obtained, and 100-fold amplifications are very readily available. Used as an oscillator, the dynatron has shown itself capable so far of producing all frequencies between less than one cycle per second and 20,000,000 cycles per second (corresponding to a wave-length of 15 meters). The output of a single bulb has been as much as 100 watts.

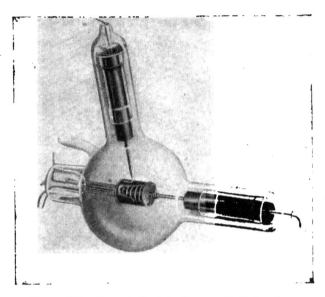

FIGURE 100—General Electric Company-Hull dynatron,
showing internal structure.

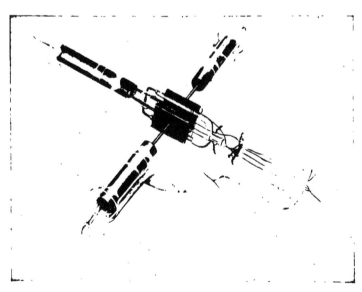

FIGURE 101—General Electric Company-Hull pliodynatron controlled
oscillator.

A still more recent device, also due to Dr. Hull, is the pliodynatron, a combination of the pliotron and the dynatron. This device has a true grid as well as the anode and plate electrodes and is an interesting four-electrode device. The grid, as usual, is an electrostatic control member, and, if the conditions are properly chosen, enables the stable control of the oscillating energy in the circuit, LC. That is, the variation of the grid potential (as determined by the battery, B'', or otherwise) will cause variations in the oscillation output of the bulb. This feature will be further considered under "Modulation Control for Radio Telephony," page 175. The wiring of a pliodynatron is clearly indicated in Figure 99.

The actual appearance of the dynatron is illustrated in Figure 100 and of the pliodynatron in Figure 101. It will be noted that the anodes are naturally much heavier than the grids of pliotrons, which must, of course, be the case, since their functions are quite different and since they must carry very considerable currents in their own circuits and be subjected to energetic electron bombardment.

CHAPTER V.

(d) ALTERNATORS OF RADIO FREQUENCY.

As we have repeatedly seen, the first necessity in radio telephony is a steady stream of alternating current of radio frequency, available for modulation into speech form. We have treated in succession the arc, radio frequent spark, and vacuum tube generators of such currents (or first approximations to such currents). It would seem, at first sight, as if we had neglected deliberately an apparently far more natural and simple means of securing such currents and one well known to ordinary commercial electrical engineering. We refer, of course, to the normal alternator.

As a matter of fact, we have deferred the study of the radio frequency alternator because of the real difficulties in the direct generation of such very high frequency alternating currents. This will be seen if we consider the pitch or distance between adjacent armature windings for a 100,000-cycle alternator. If we assume the diameter of the rotor to be 2.0 feet (60 cm.) and a normal speed of rotation of 2,500 revolutions per minute, we find that the pole pitch has the extraordinarily small value of 0.016 inch (0.04 cm.), which is entirely impracticable when one considers that wire and insulation must all be crowded into the winding slot. In addition, there would have to be 4,800 poles.

It becomes necessary, then, if we persist in the process of direct generation of the current, to have a higher speed of rotation, since the pole

103

number must obviously be reduced. Suppose we choose the extremely high speed of rotation of 20,000 revolutions per minute. We shall need then 600 poles, and the width of winding becomes 0.12 inch (0.30 cm.) approximately. So close a winding can be accomplished if great care is exercised in the choice of wire insulation and in the milling out of the slots. The requirement of a speed of rotation of 20,000 revolutions per second makes a solid steel rotor and an alternator of the inductor type essential; and this is indeed the case for the radio frequency alternators of the present, which (with the exception of the Goldschmidt type, which must have a wound armature for electrical reasons) are all of the inductor type.

We shall see that there are thus at least three general lines of endeavor in connection with the generation of radio frequent currents by alternators. These are, firstly, the multiplication of frequency within the machine (Goldschmidt type); secondly, the multiplication of frequency outside the machine (e. g., Arco alternator of the Telefunken Company, with frequency changers), and, thirdly, the direct generation in the machine of the frequency used (Alexanderson alternator of the General Electric Company). It is interesting to note that a solution of the problem of producing currents of frequencies of the order of 50,000 cycles per second (and wave-lengths of 6,000 meters) turns out to be

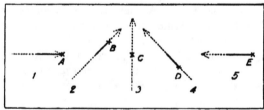

FIGURE 102—To-and-fro motion on rotating platform with equal periods of oscillation and rotation.

possible for considerable output powers (100 kilowatts or more) by all three methods. The details of these methods will be next considered.

Prior to the consideration in detail of the Goldschmidt radio frequency alternator and internal frequency changer, we desire to establish a principle of interest in connection therewith. This principle can be rendered clear from a simple analogy. Imagine a circular platform of moderate dimensions rotating once per minute, somewhat in the fashion of the carousels used in amusement resorts. Suppose, further, that the attendant elects to walk back and forth along a *diameter* of the rotating platform while it is in motion, and that he makes one to-and-fro trip in one minute, that is, in the same length of time as that required for one complete rotation of the platform. It is required to find his path as viewed from an external stationary point, or, otherwise stated, with reference to the fixed ground under the platform.

Figure 102 shows a series of successive positions of the diametral line along which he walks, each position being 45 degrees further advanced than the preceding (that is, one-eighth revolution). The dotted line with the reference dotted arrow at one end indicates this diameter which, as will be seen, has reversed its direction in the half-revolution between positions 1 and 5. The position of the man on the diametral line is indicated in each case by the cross. It will be seen that the man never succeeds in getting to the left of the center of the platform because, as position 3 is passed, he comes to the reversed end of the diametral line, that is, the end away from the arrow.

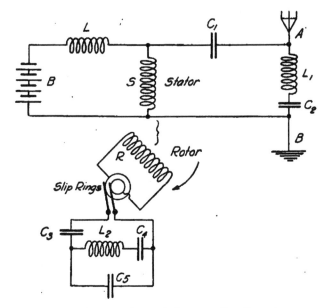

FIGURE 103—Winding of Goldschmidt alternator.

The important point is that the path of the man relative to the ground (that is, the curve $ABCDE$) is a closed curve, and that he has returned to his original position in a *half-revolution* of the platform. In other words, relative to the ground, he moves in a closed curve at twice the speed or double the frequency that the platform rotates.

We establish then the principle that an oscillatory movement of frequency n taking place on a system rotating with frequency n is equivalent relative to fixed external points to an oscillation of half the amplitude or width of swing and of *double* frequency. The mathematical proof of this principle for simple harmonic (sinusoidal) vibrations is of the utmost simplicity, but need not here be given.

The diagrammatic wiring plan of the Goldschmidt alternator is given in Figure 103. The following description is based on an earlier explanation of this device by the Author. In the figure, the battery, B, supplies the direct current whereby the stator winding, S, becomes the field magnet of the alternator. L is a large inductance intended to prevent the flow of alternating currents through the battery circuit. In the field of the stator, S, is a rotor, R, which is short-circuited (that is, tuned to resonance) for the fundamental frequency produced when the rotor is revolved. The tuning of the rotor circuit is accomplished by means of the capacities, C_3 and C_4, and the inductance, L_2. It is to be noted that R and C_3 alone would be in resonance to the fundamental frequency, as also would L_2 and C_4. The complete circuit, $R\ C_3\ L_2\ C_4$, therefore contains approximately twice the inductance and half the capacity of either $R\ C_3$ or $L_2\ C_4$. Its period, therefore, is the same as that of either of these, and even if $L_2\ C_4$ were to be short-circuited, the rotor would still be resonant to the fundamental frequency. This permits shunting the condenser, C_5, across the circuit, $L_2\ C_4$, without disturbing the tuning. A perfectly similar arrangement is adopted for the stator by the use of the circuit, $S\ C_1\ L_1\ C_2$, except that the circuit in question is tuned to *twice* the fundamental frequency. It will be seen that as the rotor revolves in the field of the stator, powerful currents of the fundamental frequency will flow through it. The great magnitude of these currents is due to the fact that the rotor is itself part of a circuit resonant to the fundamental frequency. If we consider the field of the rotor, we see that it is a field produced by an alternating current of fundamental frequency n itself rotating with a frequency, n. Therefore, by the principle established at the beginning of this discussion, we may regard it as containing a component field of constant magnitude, but rotating with a doubled frequency, $2n$, relative to the stator. A further study of the phenomena would show that there was also present a constant field rotating with velocity 0. The rotor fields will therefore induce in the stator electromotive forces of twice the fundamental frequency (and zero frequency); and since a circuit resonant to the double frequency is provided, powerful currents of that frequency will flow through the stator. These alternating currents in the stator will induce in the rotor electromotive forces of frequencies, n, (from the steady field) and $3n$ (from the field of the current of frequency $2n$). By means of the condenser, C_5, a path resonant to the frequency, $3n$ is provided in the rotor. By properly choosing the constants of the rotor circuits, the current of frequency n just mentioned can be made nearly to neutralise the current of frequency n first mentioned. The reason for this is that these currents can be brought to nearly complete opposition in phase and equal amplitude. There will be left then in the rotor a powerful current of triple frequency. Its field may be regarded, by a process of reasoning quite similar to that originally employed, as equivalent to two constant and equal rotating fields, revolving in opposite directions, with speeds of rotation corresponding to $2n$ and $4n$. There will, therefore, be induced in the stator currents of frequency $2n$ and $4n$. Of these, the current of frequency $2n$ will nearly completely neutralise the current of frequency $2n$ mentioned previously if the stator constants are properly chosen. The outstanding current of frequency $4n$ is shown in the figure as flowing into the capacity and inductance formed by the antenna, A, and the ground, B. We have, therefore, by "internal reflection" of energy, quadrupled the original frequency of the machine before using it for antenna excitation.

In the actual Goldschmidt installations (at Tuckerton, New Jersey, and Eilvese, Germany,) the motor drive of the alternator is accomplished by a 220-volt, direct current, 250-horse power motor having a speed of 4,000 R. P. M. For constant speed, a special form of sending key is used. This is shown in Figure 104. This key automatically inserts (by opening

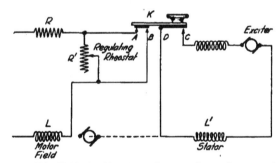

FIGURE 104—Goldschmidt alternator speed constancy system.

the back stop circuit) the resistance, R', in the motor field circuit just before the load is thrown on by closing the exciter circuit of the alternator (by the front contact of the key). In this way the motor tends to speed up just as load is thrown on, and the speed actually remains constant. In addition, the inertia of the heavy armature helps greatly.

The alternator itself is a 360-pole machine having a pole pitch or distance between windings of 7.5 mm. (0.3 inch), the slots in which the insulation and wire are placed being circular and of cross sectional

diameter of 5 mm. (0.2 inch). The rotor diameter is, therefore, **about**
90 cm. (3 feet) and the rotor weighs about 5 tons (4,500 kg.). **The**
direct current power required for field excitation is about 5 per cent. **the**
rated output of the machine.

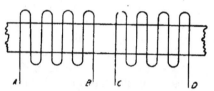

FIGURE 105—Portion ˙of rotor or stator
winding of Goldschmidt alternator
(developed).

The winding of the machine is one conductor per pole, being a
simple wave winding indicated in Figure 105. *AB* and *CD* are typical
separate sections of the winding so arranged that they may be connected
in series or parallel, depending on ,the electrical requirements. There
are twenty-four such sections on the total circumference. Both rotor and
stator are wound in the same way. The wire itself is very finely stranded,
and made of number 40 Brown and Sharpe gauge* individual enamelled
wires suitably twisted. The iron in the machine is very finely laminated,
the sheets being only 0.05 mm. (0.002 inch) thick, insulated by paper
between, 0.03 mm. (0.001 inch) thick. The rotor is more than one-third
paper, which is a most unusual proportion. Such construction is particularly noteworthy in view of the high speed of peripheral rotation,
namely 200 meters (600 feet) per second. The design of the brushes,
bearing on the rotor slip rings, and the connection to these brushes required careful consideration, especially in view of the danger of burning
the slip rings of any brush that was connected to an output circuit of
greater or less impedance than any of the remaining circuits. In this
connection, it must be mentioned that there were really more than one
pair of slip ring connections to the rotor since a number of the rotor
sections were placed in parallel outside the machine.

Some difficulty was experienced in preventing the currents which
were generated from escaping to ground through the capacity (in air)
between the conducting wires and the ground. In addition, there was
always the danger that this air capacity would, in conjunction with the
inductance of one or more of the machine windings, produce a circuit
resonant to one of the frequencies generated, whereupon dangerously
high voltages and currents would have arisen, and the output have disap-
peared.

*Diameter of number 40 wire=0.0031 inch=0.079 mm.

The accuracy of construction of such machines is extreme. Since the air gap clearance between rotor and stator is 0.8 mm. (about 0.03 inch), very accurate centering of the rotor was necessary. In addition, very strict parallelism of the armature and stator slots was required, a deviation from parallelism of one part in a thousand causing a fifth of the output of the machine to disappear!

One of the Goldschmidt alternators in use at Eilvese (Hanover, Germany), is shown in Figure 106. The machine is to the right, the

FIGURE 106—Goldschmidt alternator, motor, and reflection circuits at Eilvese, Germany.

driving motor to the left. The large brush surface chosen for the high-speed driving motor is significant. The condenser banks for tuning the various rotor and stator circuits are mounted on the walls, and are typical mica condensers. Some idea of the difficulty of leading the radio frequency currents into and out of the machine may be gained from the leads which are visible. The ingenious fashion in which the difficulties have been overcome is worthy of comment.

By January, 1917, two such alternators were being used in parallel when necessary, and put 275 amperes into the Eilvese antenna. Rapid telegraphy at a rate of 200 letters per minute has been accomplished by their use.

As has been previously stated, the second method of securing considerable amounts of sustained energy at radio frequencies when using alternators is that wherein an alternator of moderately high frequency is employed and the frequency is multiplied by external frequency changers and not, as in the Goldschmidt machine, by reflection of the energy in the machine itself. Most of the external frequency changers employed at the present time, particularly for considerable energy, are based on the properties of iron. Before explaining them in detail, it is desirable to quote from a paper by the Author on the subject of "Radio Frequency Changers."

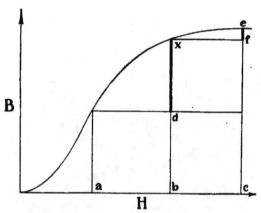

FIGURE 107—Magnetising force and magnetic
induction curve for iron.

In Figure 107 is shown a typical "B-H" curve for iron. This is the curve which shows the connection between the magnetising force (e. g.. expressed in ampere-turns or product of current flowing through the magnetising winding by the number of turns of winding) and resulting magnetisation or magnetic flux through the iron core (referred to as the "induction"). "Let us suppose that the magnetisation of the iron has been brought to the point, x. If now, by means of a superposed alternating magnetising force (such as may be produced by having around the iron core an auxiliary winding through which flows alternating current), equal increments and decrements be added to the magnetising force, the magnetic induction will increase during the positive half of the cycle by the small amount, ef. On the other hand, during the negative half of the cycle the induction will diminish by the considerably larger amount, xd. The explanation of this phenomenon is found in the well-known magnetic saturation qualities of iron, whence it results that for high magnetising forces the iron becomes saturated and the bend or "knee" of the curve which is shown at x results. It will be seen, then, that though a sine-wave alternating current may be flowing through the auxiliary winding, the variation in the magnetic flux through the iron core will not be sinusoidal but distorted, the upper halves of the curve being flattened. Such a deformation of the flux variation always occurs when nearly saturated iron cores are used under the conditions mentioned. However, such a deformation of a sine curve always leads to the production of upper harmonics (i. e., high frequencies in a secondary circuit wound around the same iron core), and it is upon this principle that the entire series of frequency changers employing iron is based."

An application of the principle just stated was shown by Epstein in 1902 (German patent 149,761) and has since been worked out and amplified in detail by Joly in 1910 and Vallauri in 1911. It is now extensively employed in various forms by the Telefunken Company under the patents of Count von Arco and Dr. A. Meissner. The circuit arrangement in a simple form is shown in Figure 108. As will be seen, an

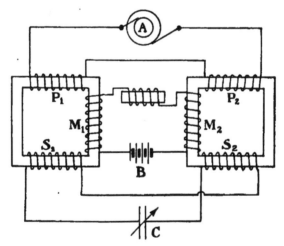

Figure 108—Telefunken Company frequency doubler.

alternating current source, A, sends its current through the primaries, P_1 and P_2, of each of two transformers having iron cores. These primaries may be connected in series or in parallel according to the secondary voltage and primary current which may be desired. They are wound oppositely relative to each other. A direct current source, B, e. g., a storage battery or small direct current generator, supplies the two auxiliary coils, M_1 and M_2, which coils are also wound on the same transformer cores. The direct current coils are wound oppositely. The secondaries of the two transformers, S_1 and S_2, are wound in the same direction, and connected as indicated in the figure.

The operation of the device is in the main as follows: The direct current flowing through M_1 and M_2 is so chosen that the iron is brought to the knee of the magnetisation curve, i. e., the point, x, in Figure 107. In consequence, during half the alternating current cycle, each of the transformers has a flattened addition to its iron magnetisation due to the iron saturation, while during the other half of the cycle it has a peaked diminution in its iron magnetisation due to the rapid drop of the iron curve below the point, x.

This effect is shown graphically in Figure 109. In curve a of the

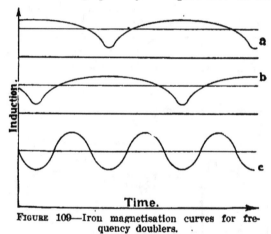

FIGURE 109—Iron magnetisation curves for frequency doublers.

figure, the fine horizontal line represents the constant magnetisation produced by the direct current which flows continuously. The curved line shows the actual magnetisation which results when the alternating current also flows in the winding, P_1. It will be seen that during the positive half of the alternating current cycle, there is only a small increase in the iron magnetisation, whereas during the negative half cycle, there is a large diminution in the iron magnetisation. It will further be noticed that the direct current coils and the alternating current coils on the two transformers are wound so that during the positive half cycle they assist each other on one transformer and that they simultaneously oppose each other on the other transformer. From this it follows that the induction in the second transformer is given by curve b, which lags practically a half cycle behind curve a. The resulting total magnetisation is given by curve c and is seen to contain a double frequency. Oscillograms of the voltages induced at the secondary terminals of each of the transformers are represented in Figure 110. The voltage at the terminal of one of

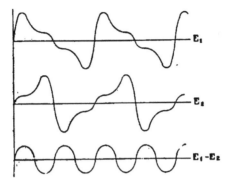

FIGURE 110—Induced voltages in second-
arys of Telefunken Company
frequency doubler.

the transformers is given by the curve E_1; that at the terminals of the
secondary of the other transformer by E_2, and there is also shown the
resultant voltage, namely E_1-E_2. The voltage curves are easily explicable
on the ground that the voltage magnitude is proportional to the rate of
change of the primary current so that it is only at times when the

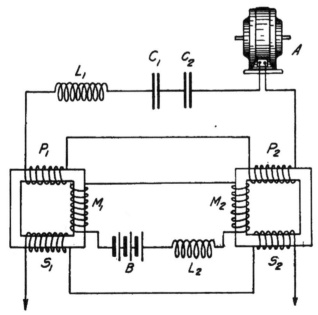

FIGURE 111—Frequency doublers used in actual practice.

primary current is changing from the flat portion to the peaked portion that the large secondary voltages are induced. The resultant voltage is seen to be of "double frequency." Its purity of wave is exaggerated in the figure.

Of course, the phenomena shown are for the frequency doubler with no load on the secondary, and these are to some degree modified when the double frequency energy is withdrawn. However, by secondary tuning and appropriate design, the same results as outlined can be obtained. A more detailed diagram, showing something of the actual practice with the frequency doublers, is given in Figure 111. It will be noted that the primary circuit of the alternator A is tuned by the inductances L_1, P_1, and P_2 and by the condensers C_1 and C_2. It will also be seen that there is a choke coil L_2 inserted in the direct current magnetising circuit of the frequency changers to prevent injuriously large radio frequency currents from being induced in this circuit.

It can further be shown, both theoretically and practically, that if the secondaries of the frequency changer, S_1 and S_2 are connected assist-

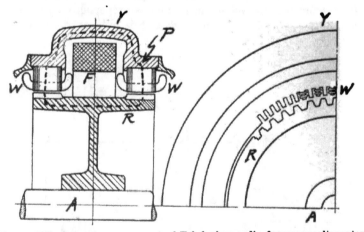

FIGURE 112—General arrangement of Telefunken radio frequency alternator.

ing instead of opposing each other, there will be produced in the secondary circuit an electromotive force of *triple* frequency. Thus the same equipment can be readily used either as a doubler or a tripler.

A clear idea of the interior construction of the Telefunken radio frequency alternators can be obtained from Figure 112. The left hand portion of the figure gives a vertical cross section of half of the machine. Here A is the shaft to which the driving motor or engine is attached either directly or through appropriate gearing. R is the inductor or rotor, a rotating mass of steel, on the outer surface of which are cut a great number of grooves parallel to A thus producing the longitudinal teeth and slots indicated in cross section at R in the right hand portion of the figure. The constant direct current passing through the field winding, F (which is an ordinary circular coil or ring of square cross section), produces a field the lines of force of which take the path indicated by the dashed line, P. It will be seen that this path is suitably interlinked with the coil, F, and passes through the yoke, Y, the stator slot supports W, and the rotor, R. The armature, which is in two portions, one on each

FIGURE 113—Telefunken Company 10 K. W., 10,000 cycle alternator.

side of the field coil consists of to-and-fro windings in longitudinal slots parallel to those of the rotor. The portions of the armature can be placed in series or parallel in accordance with the characteristics of the circuit to which the machine is connected. The mode of winding the armature is indicated at W in the right hand portion of the figure. It is evident that as the rotor revolves, the field passing through the armature turns, W pulsates back and forth with a frequency corresponding to the product of the number of rotor slots and the rotor revolutions per second. The advantage of this (inductor) type of machine as compared to those with wound armatures is that the rotating portion consists of a solid steel mass and is consequently much more sturdy than a normal armature carrying wire windings on a laminated support.

The appearance of a small (10 K. W.) machine of this type is indicated in Figure 113. The motor is mounted at the front of the base plate and the alternator at the rear. The housing between them contains the multiplying gear. The motor starter, and the speed controlling rheostat are mounted on the wall at the rear. The machine shown produced 10,000 cycles per second directly. Its use in radio telephony, together with the

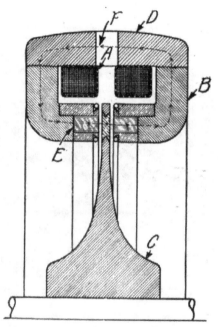

FIGURE 114—General Electric Company—
Alexanderson alternator.

other portions of a frequency changer set of which it was a part, will be described under "Control Systems," on page 190.

Continuing our consideration of the generation of radio frequency currents by alternators, we pass to an interesting and important form of alternator largely developed by Mr. E. F. W. Alexanderson of the General Electric Company. This machine has generally been distinguished by the direct generation of the very high frequency desired, and its construction has given rise to numerous difficult problems. The experimental work in connection with these alternators was originally undertaken by the General Electric Company at the suggestion of Mr. R. A. Fessenden, then associated with the National Electric Signaling Company; and much of the earlier development was done in conjunction with that company.

In 1908, Mr. Alexanderson described a 100,000 cycle alternator of this type, built to deliver approximately 2 kilowatts. A later description given by him follows (with brief additions by the Author):

"The alternator is of the inductor type (that is, with stationary armature and field, but with a rotating element which causes a pulsating field to cut the armature conductors), and is provided with a novel arrangement of the magnetic circuit, allowing the construction of a rotor which can be operated at exceedingly high speeds. In the final form of the alternators, shown in Figure 114, the rotor, C, consists of a steel disc with a thin rim and much thicker hub, shaped for maximum strength

FIGURE 115—Rotor and shaft of 100,000 cycle Alexanderson alternator.

(that is, with a width that progressively diminishes from the shaft out, so that the outward strain on the material because of centrifugal force is the same from the shaft to outer rim). The field excitation is provided by two coils, A, located concentric with the disc and creating a magnetic field the lines of force, F, of which pass through the cast iron frame, D, the laminated armature support, B and E, with its teeth, and the disc, C. This flux also passes through the narrow air gaps on each side of the disc rotor, and is indicated in the figure by the dashed line with arrows. B represents the two armatures which are secured in the frame by means of a thread, in order to allow an adjustment of the air-gap, the laminations carrying the armature conductors being located at E. Instead of poles or teeth, the disc, C, is provided with slots which are milled through the thin rim so as to leave spokes of steel between the slots. The slots are filled with a non-magnetic material (phosphor bronze) which is riveted in place solidly, in order to stand the centrifugal force and to provide a smooth surface on the disc so as to reduce air friction. The centrifugal force on each slot filler is no less than eighty pounds (37 kg.) at the high speed at which the machine is run.

"The standard 100,000 cycle rotor of chrome nickel steel with 300 slots is shown in Figure 115." The shaft bearings are clearly visible at the ends, and it will be seen that they are arranged so as to make forced oiling practicable. The shaft in this type of alternator is long and flexible, thus permitting the rotor to center itself and rotate about its center of mass somewhat as is done in the case of centrifugal dryers for laundries. In this way, excessive shaft strains are avoided. There are certain speeds (1,700 and 9,000 R. P. M.) for which the shaft and rotor pass through their own resonant periods of mechanical vibration, and at these speeds marked shaft vibration tends to occur.

A closer view of a portion of the rotor, showing the slot fillers of non-magnetic material, is given in Figure 116. Some idea of the care

FIGURE 116—Portion of rotor of Alexanderson 100,000 cycle alternator, showing slots in disc.

required in the construction of such a machine can be gained from the details of the rotor construction. Since the speed of rotation of the rotor is 20,000 revolutions per minute, or over 330 revolutions per second, the actual speed at the rim is nearly twelve miles per minute! Such a machine must, accordingly, be considered a masterpiece of engineering design. (A whimsical calculation has been made which shows that the rotor, if released while spinning at full speed, would, if it maintained its speed thereafter, roll from America to Europe in a few hours!)

There are two methods of armature winding employed in the simpler forms of these machines. The first form, which is a simple to-and-fro winding (one turn per slot) is shown in Figure 117. In this form of armature there are 600 slots for a 100,000 cycle machine. A second form of winding for the armature has only 400 slots for the 100,000 cycle machine. It is shown in Figure 118, and really consists of two windings

FIGURE 117—Portion of armature winding of 100,000 cycle Alexanderson alternator; 600-slot type.

in parallel in each of which, by a sort of vernier action, a 300-slot rotor field produces 100,000 cycle current in the same phase in each of the armature windings. *It is possible, using an 800-slot armature winding of the last-mentioned type, to produce a 200,000 cycle current by direct*

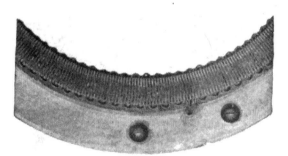

FIGURE 118—Portion of armature winding of 100,000 cycle Alexanderson alternator; 400-slot type.

generation. This is by far the highest frequency which has as yet been produced directly by an alternator.

Through the courtesy of John L. Hogan, Jr., of the National Electric Signaling Company, we are enabled to show in Figure 119 a test of an early form of 80,000 cycle alternator built by the General Electric Company and used at the Brant Rock station of the National Electric Signaling Company in 1906. This machine had a double inductor with inward projecting teeth on each half, and the stator lay between the two "saucer" shaped inductors. It will be seen that this machine was belt driven to get the proper ratio of motor to alternator speeds, and that the motor is much larger than the alternator. This is quite explicable when it is remembered that the windage loss in these machines at 20,000 R. P. M. is high, it having been claimed that the rotor is actually polished either by air friction or by the friction of floating dust particles. In any case, the

FIGURE 119—Early form of Alexanderson alternator under test
at Brant Rock Station of National Electric
Signaling Company.

air streaming out from the machine is appreciably warmed. This windage loss becomes important in any but the smallest alternators of this type.

A somewhat similar machine built by the National Electric Signaling Company in 1907 and equipped with de Laval steam turbine drive is shown in Figure 120. This has the advantage that, since the turbine is itself an extremely high speed machine, the gearing losses are eliminated by the direct drive. Sufficiently accurate speed regulation of a steam

FIGURE 120—Early form of Alexanderson alternator coupled to
de Laval turbine; under test by National
Electric Signaling Company.

driven machine is secured in practice by maintaining the steam pressure
and radio frequency load at constant values. The gearing shown in the
figure is used to reduce the main shaft speed in the ratio of 1-to-10 for
the operation of the turbine governor. It will be noted that the alternator in this figure has an adjustment to rotate each armature slightly
relative to the frame so as to bring the generated currents into phase and
also has an adjustment whereby, as stated previously, the armatures may
be brought nearer to or further from the rotor for precise adjustment of
the air gap. Such an adjustment is of importance since the output of

FIGURE 121—Intermediate type of Alexanderson alternator.

the machine is largely dependent on the air gap, and a very small air gap (of 5 or 10 thousandths of an inch, or an eighth to a quarter of a millimeter) is of advantage. The usual gap is 0.015 inch (0.38 mm.) with a generated voltage of 150, although voltages as high as 300 can be obtained with a 0.004 inch gap.

This machine was in almost daily use at Brant Rock for several years, and ran for hours at a time without attention. The maximum output was something over 1 K. W. at 100,000 cycles.

A later form of a 2 K. W. Alexanderson alternator is shown in Figure 121. This set shows the elaborate forced-feed oiling system which has been adopted for the later machines. The auxiliary and main bearings to the right of the rotor are clearly visible.

The most recent form of 2 K. W. machine of this type is shown in Figure 122. The oiling system in this machine is provided with an inter-

FIGURE 122—Recent type of 2 K.V.A., 100,000 cycle Alexanderson alternator.

esting protective device. The oil which is returned to the reservoir at the right of the base plate (the tank having a sheet metal cover with handle) strikes a small pivoted shovel. Its weight depresses this shovel against a controlling spring tension. Should the flow of oil cease for any reason, the shovel flies up and automatically opens the driving motor circuits. In this way, any danger of unoiled bearings "freezing" is obviated. In this set, the alternator is driven by a 110 or 220-volt, direct current, shunt motor with commutating poles. The motor speed is 2,000 revolutions per minute and this is raised to the requisite 20,000 revolutions per minute by the 1-to-10 helical-cut gearing enclosed in the housing at the center of the base. The oil pump, which is chain driven from the motor shaft, is shown at the right hand corner of the base. To prevent any possibility of binding between the two thrust bearings, due to expansion of the shaft because of heating, the machine is provided with a system of equalizing levers to compensate for such shaft heating. These levers are shown in the left front of Figure 122 with the elastic controlling leaf between them. Any tendency which would cause a change in air gap is counteracted by the automatic action of the levers. If the air gap should tend to change at either side, the magnetic attraction at that side would cause an additional pressure and consequent heating on the thrust bearings at that end; and a consequent expansion of the shaft there would bring the rotating disc back to a central position.

The expansion of the shaft by temperature is thus taken advantage of to insure a correct alignment. The usual output of these alternators is from 10 amperes at 200 volts to 20 amperes at 100 volts, depending on the nature of the load and the mode of internal connection of the armature sections of the machine. The effective resistance of the armature is 1.2 ohms, the inductance being 8.6 microhenrys corresponding to 5.4 ohms, at a frequency of 100,000 cycles, or wave length of 3,000 meters. The resonance condenser load would, therefore, be 0.29 microfarad at the frequency mentioned, if no loading coil were used external to the machine.

Another recent type of Alexanderson radio frequency alternator is the so-called "gyro alternator." The designation is based on the similarity of bearings in the machine in question and those in a high speed gyroscopic compass. A heavy shaft is used, so that vibration at the "critical speeds" does not occur, these speeds being much higher than those at which the machine is actually run. The use of ball bearings in this machine has simplified the construction. No auxiliary bearings are needed in this machine.

FIGURE 123—Recent General Electric Company-Alexanderson alternator of "gyro" type.

Figure 123 shows one of these machines with belted driving motor and all auxiliaries needed for a complete radiophone equipment mounted on a base. The particular equipment shown has been used under favorable conditions for the transmission of speech 160 miles (250 km.), between Schenectady and the Author's laboratory in New York. The alternator generates 33,000 cycles per second, which is transformed into 100,000 cycles (corresponding to a wave length of 3,000 m.). The 100,-000 cycle energy is modulated by a magnetic amplifier which is controlled directly by a standard microphone. A description of the magnetic amplifier system of modulation follows under "Modulation Control Systems," page 195.

Passing from the smaller machines, Mr. Alexanderson has had built a 50 kilowatt, 50,000 cycle alternator (and very considerably larger machines are under test and construction). This machine is shown in Figure 124. The open circuit voltage of this machine and the trans-

FIGURE 124—50 kilowatt, 50,000 cycle, General Electric Company-Alexanderson alternator.

former described below is about 550 volts, but the machine is normally operated at about 125 amperes and 400 volts. The rotor is similar to, although naturally larger than that of the smaller machines previously described, but an extremely heavy and rigid shaft is used. The machine has proven capable of furnishing 85 kilowatts for brief periods. Operating at 3,500 revolutions per minute, its bearings and shaft construction are similar to those of normal high speed turbines. The machine speed never attains the "critical speed" value, thus avoiding the necessity for auxiliary bearings. Because of the very rigid shaft, the rotor is not measurably deflected by the magnetic field. "The thrust bearings for

the collars shown at each end of the rotor shaft are held in position with
a system of equalizers, which have for their object the avoidance of any
possibility of binding in the bearings due to expansion of the shaft from
change in temperature, and at the same time automatically draw up all
slack in the bearings as they become worn. The equalizers are the heavy
vertical columns and links shown in the photograph of the assembled
machine.

"The direct generation of radio frequencies by a machine working
on the principle of a simple alternator is possible only by the use of a
very low voltage winding. On the other hand, if the alternator windings were designed to be connected directly in series with an antenna, the
terminal voltage would be about 2,000 to 3,000 volts. Thus it is apparent
that with this type of machine it is necessary to use a transformer between the machine and its output circuit. The alternator windings con-
sist of thirty-two independent circuits connected to the same number of
independent primaries of the transformer. The transformer has a
number of secondary circuits which can be connected for various ratios
of transformation between 4-to-1 and 24-to-1. Thus the alternator can
be adapted to antennas of greatly different characteristics. The primary
windings of the transformer are grounded in the middle, so that the
greatest potential difference to ground on the alternator winding is one-
half the voltage generated by one alternator circuit.

"The transformer is a closely coupled one, the coupling coefficient
being 0.95. In the phraseology of the alternating current designers, the
transformer may be described as having about 30 per cent. magnetizing
current and 30 per cent. total leakage. Although the transformer has
no iron core, it has a measurable core loss due to the eddy currents in the
conductors caused by the magnetic flux. If it were not for these eddy
currents, the efficiency of the transformer would be close to 99 per cent.;
as it is, the efficiency is about 95 per cent. This efficiency is approximately constant between frequencies of 25,000 and 50,000 cycles, because what the transformer in one sense gains by the higher frequency,
it loses on account of the higher eddy current accompanying that frequency. The numerous multiple circuits in the primary, as well as those
in the secondary, are carefully transposed so as to make cross currents
impossible between the different circuits.

"While it appears that the most practical arrangement from all
points of view is the one described, i. e., a low voltage winding and transformer, experiments have been made with windings distributed in such a
way that larger slots can be used with room for more insulation. A sample
machine of this type of 3 k.w. output at 45,000 cycles was built, and a
diagrammatic representation of the armature cross section and rotor is
given in Figure 125. This generates a frequency three times as high as

the one for which the slots on the winding are apparently designed. This method may be characterized as generating triple harmonics without the fundamental. The action is somewhat like that of a vernier, the flux through the stator projections changing from that due to two teeth on the rotor to that due to one tooth at three times the apparent frequency of the machine. While the characteristics of this machine have proven entirely satisfactory, in accordance with expectations, it is probable that the original simple form of winding will be adhered to, because the concentration of large conductors with more current in one slot causes not only higher losses, but also a lower rate of heat dissipation and therefore less output can be expected from the same amount of material.''

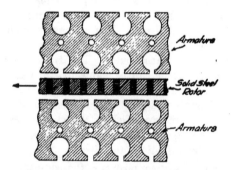

FIGURE 125—Diagrammatic representation
of Alexanderson alternator for direct
generator of triple frequency.

It may here be mentioned that the machines shown in Figures 121, 122, 123, and 124 have all been used for radio telephony in connection with further devices which will be described under ''Control Systems.'' The first was used principally by the National Electric Signaling Company in Mr. Fessenden's tests between Boston and New York (Jamaica), a distance of some 150 miles (240 km.). This was, however, not a matter of regular communication, but rather of test work. The machine shown in Figure 122 has enabled quite regular communication between Schenectady and New York, the distance being 150 miles (250 km.). Even the smallest machine (of Figure 123) running on much reduced power, has enabled the same stretch to be bridged when suitable receiving apparatus was employed.

With the large machine shown in Figure 124, employing the magnetic amplifier controlling device to be described hereafter, the output was successfully modulated between 5.8 kilowatts minimum and 42.7 kilowatts maximum. This is, to date, the maximum amount of radio frequency energy controlled telephonically by any means.

CHAPTER VI.

7. MODULATION CONTROL IN RADIO TELEPHONY.

We have, up to this point, considered many matters which are common to radio telephony and radio telegraphy since the sustained wave generating systems are naturally applicable to the latter field as well as the former and indeed were originated principally in connection with telegraphy. We pass now, however, to a matter exclusively related to radio telephony, namely the modulation or control of amounts of power varying from a few watts to many kilowatts *by the human voice.* The problem is indeed a difficult one, and for a long time practically defied solution. When it is considered that the rate of energy radiation in the form of sound in ordinary speech is of the order of one one-hundred-millionth to one-billionth (0.000 000 01 to 0.000 000 001) of a watt and that the delicate and excessively complex variations of the sound energy must be faithfully reproduced with an energy amplification of *hundreds of billions,* and that the energy to be modulated is of the peculiar form associated with radio frequency currents, the difficulties of the problem become evident. And yet radio telephony is entirely dependent on the simple solution thereof.

Before describing in detail the various methods of modulation control which have been devised, we shall consider certain broader questions connected therewith. The first of these is the completeness of control systems.

(a) DEGREE OF CONTROL.

In every radiophone transmitter, there is some point at which a controlling current, voltage, inductance, capacity, or resistance exists. The current might be, for example, the fluctuating current in a telephone transmitter circuit. The voltage might be the voltage applied to the grid

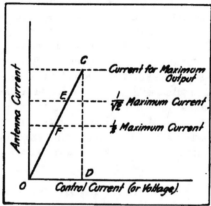

FIGURE 126—Complete linear modulation control curve.

of a vacuum tube of some sort, this voltage being derived from the secondary circuit of a transformer connected in a microphone transmitter circuit. The resistance might be the resistance of a microphone placed directly in the antenna of a radiophone transmitter. Whatever the controlling element, it must vary between certain extreme values when speech is being transmitted, these limits being reached for the peaks of the loudest sounds which are encountered in ordinary speech. Indeed, it is preferable that these peaks should be reached for such normal speech rather than for shouting since otherwise the control tends to be weak and excessive amplifications may be required somewhere in the set.

In Figure 126 we have a control characteristic of a desirable sort. Horizontally is plotted the controlling element (current, voltage, capacity, inductance, resistance, or a combination of these), this element varying between zero and the value $O\ D$. Vertically are plotted values of the controlled or antenna radio frequency current, this varying between zero and *CD*. It will be seen that we have a straight line characteristic curve; that is, the controlled current is proportional to the controlling current. Furthermore, the control is complete, since we assume that the current *CD*, is the greatest current which the sustained generator is capable of putting into the antenna; or, in other words, the current *CD* is determined by the actual maximum possible output of the alternator, arc, radio frequent spark transmitter, or vacuum tube transmitter used.

The question arises as to what will be the ammeter reading in the antenna when no speech is being sent out. This depends on whether we choose to have this point as that of half current or of half energy. In the former case, the current will center around the point *F* which represents one-half the maximum current. In the latter case, the current will rise and fall about the point *E* which represents the reciprocal of the square root of two (that is, 0.707) times the maximum current. Under normal conditions of speech, in some types of radiophone (particularly those with stable control to be described hereafter) the antenna ammeter does not change markedly from the point *E* (or *F*) when one actually carries on conversation. In other types (and especially those with unstable control) the average current in the antenna may change considerably when speech is being transmitted.

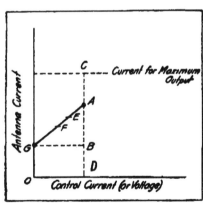

A curve representing incomplete control is given in Figure 127. It will be noted that the entire available variation in the controlling element will cause a variation in the antenna current only between the values *OG* and *DA* and not between zero and the maximum available current *CD*. In this case we may define the percentage of control as the quotient of *AB* divided by *CD*. Such a radiophone set with a normal linearly proportional receiver will be equivalent to a considerably less powerful transmitter than that represented in Figure 126.

FIGURE 127—Incomplete linear modulation control curve.

The Author advocates control characteristics in which the microphone current is taken as the controlling element and the antenna current as the controlled element. First of all, these elements are fairly readily measurable. In the second place, what is, after all, desired is that the *current* variations through the receiver telephones shall be proportional to the *current* variations in the microphone transmitter as in an ordinary telephone line. It is accordingly deemed best to adhere to current control characteristics throughout.

In practice, the perfect type of control characteristic shown in Figure 126 is not realized. A more common form which is fairly acceptable

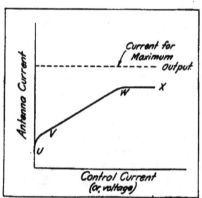

FIGURE 128—Typical incomplete non-linear modulation control curve.

is shown in Figure 128. It will be seen that the antenna current never falls below U although this leads to a waste of energy. From V to W the control is linear and satisfactory, but at W it flattens, remaining at constant current to X. The maximum output current is never reached, the mere existence of the controlling element preventing its attainment.

(b) STABILITY OF CONTROL.

The control system of a radiophone may be classified as stable or unstable depending on whether the points on the upper and lower portions of the control curve can be held steadily with the control system used or whether they can be reached for only a brief period of time. The simplest example of a stable control system would be the following: Imagine a radio frequency alternator, driven by a constant speed motor, placed directly in a tuned antenna; and in series with the alternator and directly in the antenna a microphone transmitter (or a variable resistance, which is its equivalent). It is evident that the system is perfectly stable no matter what the value of the microphone resistance, since the only possible effect of an increase or diminution of the microphone resistance is to lower or raise the antenna current. The control curve of such a system would be

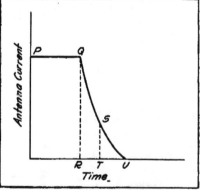

FIGURE 129—Dynamic characteristic of falling current for unstable control system.

a "static" characteristic; i. e., one for stationary conditions. The simplest example of an unstable control system would be the following: A Poulsen arc is placed directly in the antenna in series with a microphone (or a variable resistance). Changing the resistance of the microphone will not merely cause the antenna current to change; it may actually cause the extinction of the arc altogether, if the inserted resistance is too high. So that the system would be unstable at this limit. This is illustrated in Figure 129 where the antenna circuit is plotted against time. The current remains constant from P to Q, it being supposed that the resistance in the antenna is moderate for this range of time. At Q a large resistance is inserted in the antenna, for example, by speaking into the microphone, or pulling out its diaphragm. The antenna current may not merely decrease; it may rapidly fall to zero at the point U. On the other hand, if the resistance is restored to its former small value after a time RT, the antenna current may recover the full value Q rapidly. In other words, while the system is unstable for permanent changes, it may be operative for rapid transient changes provided these changes are of very short duration. For this case, it is not possible to secure a complete static control characteristic; dynamic characteristics must be obtained by the use of an oscillograph or some other device for following the rapidly changing antenna and controlling currents.

Generally speaking, unstable control systems are objectionable. If the time RU of extinction in such a system is very short, then a low-pitched sound (of relatively long period) may lead to complete "breaking" or extinction. On the other hand, if the time RU is long and the slope of QSU less abrupt, there will be sluggishness of control and a blurring or muffling of speech. Rigid and stable control is desirable.

(c) RATING OF RADIOPHONE TRANSMITTERS.

In a receiving set, when audibility measurements are being made on received speech (on the basis of just hearing sound of any character), it is the maximum transmitter current (CD of Figure 126) which is being considered. Consequently the Author recommends that radiophone transmitters be rated on the basis of maximum energy radiated, corresponding to maximum current. Here 100 per cent. control is assumed. If less than full control is attained, the rating of the transmitter should be the *maximum energy variation*. As has been stated, for unstable control systems, this requires an oscillograph for the determination of rating; but generally we may assume the maximum energy in this case to be twice the average or steady value, if the control is known to be linear. Many unstable control systems flatten out in control (as at WX in Figure 128) for high currents, and consequently their rating may be much less than that given by the double energy rule above.

In rating radiophone transmitters on the basis of maximum energy radiation, it must be understood that this does not imply that a 1 K. W. radiophone transmitter will enable the clear transmission of speech for the same distance as a 1 (antenna) K. W. spark transmitter will enable the transmission of telegraphic signals. More than just the peaks of the received speech is required for comprehensibility, so that the received speech must be considerably more than once audibility to be fully understood. The exact number of times audibility required for satisfactory speech is not precisely determined at present and depends naturally on the freedom from speech distortion. It is probably not less than 2 nor more than 10.

(d) TYPES OF CONTROL.

Control systems may further be classified as absorption systems or generator voltage (or current) control systems. The simplest instance of an absorption system is the plain microphone-in-antenna modulation where the microphone actually absorbs intermittently a considerable portion of the radio frequency generator output. Such systems, while distinguished by their simplicity and satisfactory behaviour for small powers are not so easy to apply to large powers because of the difficulty of absorbing considerable amounts of energy in any system sufficiently delicate to follow the voice inflections. Among exceptions to this statement, however, are the vacuum tube absorptive systems to be described hereafter.

The generator-voltage control type is well illustrated by the use of radio frequency alternator in the antenna, the field of the alternator being excited by the microphone current and the alternator being driven by a constant speed motor. It will be seen that the generator output is variable in this case, and not constant as in the former. This requires quite special driving motors and is an objection. On the other hand, the absorption control systems tend to be constant load systems and do not require special driving machinery. However, unless an absorption system is carefully devised, it may be uneconomical of energy, since it is desirable to avoid having full load on all machinery regardless of whether speech is being transmitted or not.

(e) MICROPHONE TRANSMITTER CONTROL.

An ordinary microphone transmitter of high resistance will carry a steady current of from 0.1 to 0.2 ampere at an applied voltage of 10 volts. Its resistance is therefore of the order of 50 to 100 ohms, and the energy which it can absorb steadily is about 2 watts. If it is attempted to pass more current than that mentioned through the microphone, a "frying" or crackling sound will be heard in the receiver, the carbon grains of the transmitter will overheat and burn, and the microphone will steadily deteriorate. A so-called "low resistance" transmitter will carry 0.4 to 0.5 ampere and have a resistance of from 10 to 20 ohms. It can absorb satisfactorily but little more energy than the high resistance form.

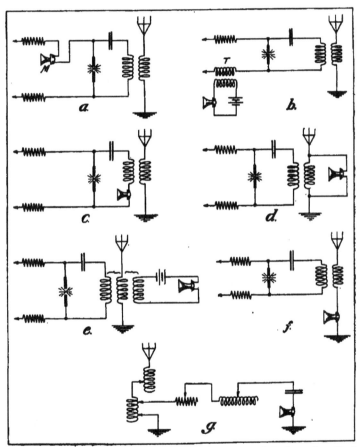

Figure 130—Various arrangements for microphone modulation of radiophone transmitter.

When a microphone overheats from the passage of excessive current, which is very likely to occur when the over-enthusiastic radiophone experimenter places it in the antenna circuit and attempts gradually to increase the antenna current, it "packs". That is, the grains of carbon expand and become tightly wedged in the carbon chamber and the microphone no longer responds. It then becomes necessary to shake the microphone mechanically to release the grains and restore its modulating power. In one form of radiophone made by Dr. de Forest, the shaking of the microphone was accomplished by fastening a buzzer to it and closing the battery circuit of the buzzer occasionally. The resulting vibration gave the desired result. A more simple means of accomplishing the same result is by tapping the transmitter. A "packed" transmitter rapidly deteriorates through overheating and burning of the carbon grains.

Dr. Georg Seibt has shown that, in order that the loudest signal shall be heard in the receiving station when a microphone transmitter is used for modulating the transmitted energy, a simple condition must be fulfilled. It is that the resistance (as determined by energy absorption of the microphone when undisturbed) shall be equal to the total resistance (as determined by energy absorption) of the remainder of the radio frequency circuits of the transmitter. For example, imagine an antenna of 8 ohms total resistance (including ohmic resistance, ground resistance, radiation resistance, and eddy current loss resistance) with an inserted microphone. The microphone resistance should also be 8 ohms. From this it is fairly obvious that a high resistance microphone is inapplicable, unless it is not in the antenna but so coupled or connected to antenna circuit (directly, inductively, or capacitively) that its effect is the same as if a smaller resistance equal to the antenna system resistance had been inserted.

We show in Figure 130 a number of arrangements which have been used for the direct control of the radiated energy by a microphone. Diagram *a* shows the microphone inserted in the direct current supply leads of the arc, thus causing appropriate variations of the arc current and arc output. Diagram *b* is somewhat different in that alternating electromotive forces are impressed on the arc as well as the constant supply voltage. The alternating voltages are transferred to the arc supply circuit through the transformer *T* connected to the microphone circuit and supply circuit. This arrangement is due to Mr. E. Ruhmer. In the Diagram *c* the microphone has been transferred to the oscillating circuit of the arc. This method would, except with very low resistance microphones, be an unstable control system. In Diagram *d*, which shows a circuit used by both Professor V. Poulsen and the Telefunken Company, the microphone is shunted across the antenna coupling and tuning coil. It would therefore act to detune the antenna circuit as well as to absorb energy intermittently. The method is quite effective. Diagram *e*, which is another arrangement due to Professor Poulsen, accomplishes the same results by coupling the microphone inductively to the antenna coupling coil. The only purpose of the battery in this case is to bring the microphone resistance (which depends on the current passing through it) to a desired value. Diagram f illustrates an arrangement used by Mr. Fessenden (principally with radio frequency alternators as generators) and others. In this simple case the microphone is directly in the antenna, and moulds the radio frequency current into the desired speech form envelope more or less fully. Diagram g shows the unusually elaborate arrangement adopted for modulation by Messrs. Colin and Jeance. A tuned circuit of desired constants is directly coupled to a portion of the antenna coupling coil. The microphone is directly inserted in the

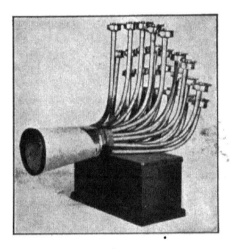

FIGURE 131—C. Lorenz Company multiple
transmitter.

tuned shunting circuit which has sometimes been characterized as a "spill-over" circuit.

In order to modulate more energy than can be properly handled by one microphone, the idea was originated of using several in series, low resistance microphones being thus employed. The idea is feasible to a limited extent, but rapidly leads to difficulties in carrying the energy of the speech to the diaphragms of many microphones. An extreme instance of this method is shown in Figure 131, which shows no less than 25 Berliner microphones being thus used by the C. Lorenz Company. A less extreme instance is illustrated in the radiophone set illustrated in Figure 21 of this book, which shows six microphones in series. It is desirable in such arrangements to have the paths from the mouth of the speaker to the different microphones of equal length and as nearly as possible geometrically identical, so that each microphone gets full excitation.

A further expedient is to have more than one set of microphones available, and to change over from one set to the next whenever considerable heating occurs. Data is not available as to the practicability of this scheme, but it seems to be of some advantage.

The multiple microphone transmitter on this principle employed by Lieutenant Ditcham is shown in Figure 132. It consists of four pairs of two microphones each, the microphones in the individual pair being simultaneously actuated by the voice and connected in series. A knob on the side of the holder (or, in some types of the apparatus, an automatic push-button arrangement) enables changing from one set of microphones to the next about every two minutes, thus preventing overheating. Antenna current up to 10 amperes have thus been handled without overheating of the microphones and consequent deterioration of articulation.

Figure 132—Ditcham multiple microphone transmitter.

Mr. R. Goldschmidt has devised a method of using several microphones in parallel. Normally this is not feasible, since if one begins to get more current than the remainder its resistance will rapidly fall and it will soon carry the entire current, thus leading to injurious overheating. The simplest form of the method mentioned is given in Figure 133. As will be seen, the microphones are each in series with a coil (L_1 and L_2 respectively). The coils in question are wound oppositely on a com-

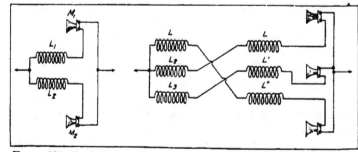

Figure 133—R. Goldschmidt's method of utilizing microphones in parallel.

mon core. As long as the current through each microphone is the same, equal currents will flow through each coil and the net inductance in the circuit will be zero. If, however, one of the microphones takes more current than the other, the balancing current begins to circulate in the circuit $L_1M_1M_2L_2$ and encounters a high inductance in L_1L_2 since it does *not* flow in the opposite direction in the two coils. The method of extending the idea to three microphones is also illustrated. Here coils L_1 and L are wound on the same core, as also are the coils L_2 and L', and the coils L_3 and L''.

(f) HIGH CURRENT MICROPHONE CONTROL.

The first thought that naturally suggests itself in connection with the securing of microphones that will modulate successfully more energy than will the ordinary carbon microphone is to replace the carbon by some more permanent and less inflammable material. Carborundum has been suggested by several inventors, but it cannot be said that any data is available favoring the belief that the expedient was successful.

In 1906 and 1907 Mr. R. A. Fessenden, at that time directing the work of the National Electric Signaling Company, devised a number of microphone transmitters which carried heavy currents for considerable periods of time. He also developed a heavy current telephone relay, which permitted controlling considerable current by means of smaller currents originating in an ordinary microphone circuit or coming from a telephone line. A description of these devices in his own words, with added comments by the Author, follows:[*]

"A new type of transmitter was therefore designed which the writer (Mr. Fessenden) has called the 'trough' transmitter. It consists of a soapstone annulus to which are clamped two plates with platinum iridium electrodes. Through a hole in the center of one plate passes a rod, attached at one end to a diaphragm and at the other to a platinum iridium spade. The two outside electrodes are water-jacketed.

"The transmitter requires no adjusting. All that is necessary is to place a teaspoonful of carbon granules in the central space. It is able to carry as much as 15 amperes continuously without the articulation falling off appreciably. It has the advantage that it never packs. The reason for this appears to be that when the carbon on one side heats and expands, the electrode is pushed over against the carbon on the other side, thus diverting a greater portion of the total current to the cooler side, which has thus been made of smaller resistance. It will be noted that the two halves of the carbon, on the opposite sides of the spade diaphragm are in parallel. These transmitters have handled

---

- "Proc. A. I. E. E.," June 29, 1908.

amounts of energy up to one-half horse power (375 watts), and under these circumstances give remarkably clear and perfect articulation and may be left in circuit for hours at a time."

Such a water-cooled microphone, built to carry up to 6 amperes continuously, and suitably mounted, is illustrated in Figure 134.

FIGURE 134—National Electric Signaling
Company-Fessenden high current
transmitter.

A more complex and extremely interesting device is shown in Figure 135. This is* "a transmitting relay for magnifying very feeble currents. It is a combination of the differential magnetic relay and the trough transmitter. An amplification of 15 times can be obtained without loss of distinctness. . . . The successful amplification depends on the use of strong forces and upon keeping the moment of inertia of the moving forces parts as small as possible. Amplification may also be obtained by mechanical means, but as a rule this method introduces

* "Proc. A. I. E. E.," June 29, 1908.

scratching noises which are very objectionable even though comparatively faint." The amplifying relay shown in Figure 135 is capable of handling 15 amperes in its output side. Thus over ten years ago Mr. Fessenden recognized the desirability of being able to control the radiophone transmitter from a wire line, and this relay was developed to enable the desired result to be obtained.

FIGURE 135—Fessenden heavy current telephone relay.

A complete radiophone station at Brant Rock embodying the idea just mentioned is illustrated in Figure 136. Although completed in 1906, the design thereof was remarkably advanced. In the right foreground are seen the radio frequency alternator and its driving motor and controlling rheostats. Directly back of these is a compressed air tuning condenser. On the table is shown a normal line telephone set connected to the high current relay which controls the outgoing energy. In addition, at the reader's left, on the table is placed a portion of the receiving set. On December 11, 1906, a demonstration of radio telephony was given from Brant Rock to Plymouth, Massachusetts, a distance of 10 miles (16 km.). Both speech and music were transmitted. In addition,

speech was transmitted over an ordinary wire line to the radio station
at Brant Rock, relayed automatically to the radiophone, transmitted by
radio to Plymouth, and at Plymouth automatically relayed back to a
wire line. Telephone experts present noted a remarkable absence of
distortion of speech quality. In July, 1907, speech was transmitted between Brant Rock and Jamaica, Long Island, a distance of 180 miles
(290 km.) over land, and by day. The antenna mast at Jamaica was
180 feet (55 m.) high. In this work, "the transmitting relays are con-

FIGURE 136—National Electric Signaling Company-Fessenden 2 K. W.
radiophone transmitter.

nected to the wire line circuit in the same way as the regular telephone
relay, except that in place of being inserted in the middle of the line, they
are placed in the radio station and an artificial line used for balancing.
There is no difficulty met with on the radio side of the apparatus, but on
the wire line there are the well-known difficulties due to unbalancing
which have not been entirely overcome. For the correction of these difficulties, therefore, we must look to the engineers of the wire telephone
companies. At present, the difficulties are, if anything, less than those
met with in relaying on wire lines alone."

Another form of transmitter used by Mr. Fessenden is the condenser transmitter. This is not a carbon microphone at all, but a variable condenser with one (or more) fixed plates and one (or more) movable plates, the movable plates being brought nearer to or further from the fixed plates by the voice vibrations. In this way there are produced in this condenser changes of capacity closely proportional to the sound amplitudes. If such a condenser transmitter be connected between a high potential point of the antenna (e. g., the topmost point of the loading coil, L of Figure 137) and ground, it will have two effects when its capacity is varied by the sound waves. To begin with, it will detune the antenna by

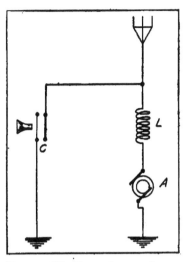

FIGURE 137—Fessenden condenser transmitter for radiophone work.

shunting the coil L and the radio frequency alternator A by a larger or smaller capacity (which capacity is, in effect, in parallel with the antenna capacity). This effect may be considerable if the antenna capacity is small, the antenna damping small, and condenser transmitter capacity variations large. Figure 138 depicts the curve of antenna current (ordinates) against frequency to which antenna is tuned (abscissas) with the alternator A run at constant frequency corresponding to the point A on the curve and the peak of the resonance curve. The proper point to work the antenna for such a system would be at some such point as F on one of the steeply falling branches of the resonance curve. Then, if the frequency were altered periodically between OB and OD by the condenser transmitter, the antenna current would similarly vary periodically between EB and DG. The second effect of varying the capacity of the condenser transmitter would be actually to "spill" energy from the antenna to ground through the transmitter capacity. These two effects should assist each other and the reader can satisfy himself by a little thought on the subject that this result can be secured by tuning the antenna system with the condenser transmitter in its undisturbed position to a *lower* frequency (that is, *longer* wave length) than that of the alternator. If the opposite is done, the two effects of the condenser may partially or entirely neutralize each other.

Continuing our consideration of high current microphones for modulation in radio telephony, we come to a type of telephone relay used by Mr. W. Dubilier. The radiophone transmitter with which it is em-

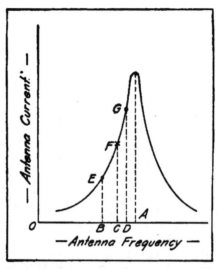

FIGURE 138—Detuning characteristic of
radiophone transmitter.

ployed has been illustrated in Figures 37, 38, and 39. The inventor (in
1911) was well aware of the advantage of transferring speech from telephone lines to the radiophone transmitter and designed the relay for that
purpose. A description thereof follows:

"Figure 139 shows a cross section of the relay. The complete transmitter consists of the magnets *A, A* wound with two-ohm winding *B, B,*

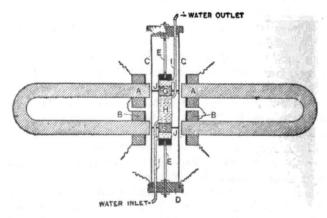

FIGURE 139—Cross section of Dubilier high current telephone relay.

and placed opposite to each other with the diaphragms and carbon containing cup between. (There had been adopted a type of transmitter with *two* diaphragms with the carbon between, both diaphragms swinging inward or outward in synchronism and thus producing greater changes in the resistance of the carbon between them than if one of them were fixed). The diaphragms are approximately 5 inches (12.7 cm.) in diameter and 0.036 inch (0.9 mm.) thick.

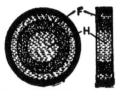

FIGURE 140—Cross section of carbon-containing cup of Dubilier high current relay.

The ebonite disc D is used to mount the diaphragms, and is drilled with large-sized holes so as to prevent 'air packing' or talking against each other.

"A cross section of the carbon-containing cup is shown in Figure 140. It resembles three brass rings placed one within another, forming three independent containing portions. Water circulates through the chambers F and H by means of the inlet and outlet tubes I, and the middle chamber G (of Figure 139) is used to retain the carbon granules. To make contact with the granular mass, circular rings of platinum, J, J are used, which are first soldered to the diaphragms C, C. The platinum rings are drilled with small holes round the entire circumference so as to allow a free circulation of air, and through one of these holes the small inlet and outlet tubes are run. The contact is made in the center of the granular mass. A mica disc is used to retain the granules in the chamber."

The transmitter in question was designed to carry currents up to 6 amperes. It seems to have been operative; since, as stated previously, radio telephony over 250 miles (400 km.) was accomplished with such apparatus.

In Figure 141 is shown in detail the high current microphone transmitter used by the Telephone Manufacturing Company (formerly J. Berliner) of Vienna. The entire

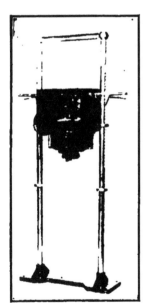

FIGURE 141—Heavy current Berliner microphone transmitter with fan cooling.

radiophone set of which it is a part was illustrated in Figure 22. The microphone is seen to be mounted on a frame support with a ratchet clutch for holding it at any desired height. On the front of its case are the large mouthpiece and a double throw switch for "Calling" or "Speaking.'' Directly beneath the transmitter is placed a horizontal fan for cooling purposes.

In building high current transmitters, the granular carbon—carbon diaphragm type (e. g., as built by the Berliner Company) has been found to be suitable. The usual modifications made therein when used

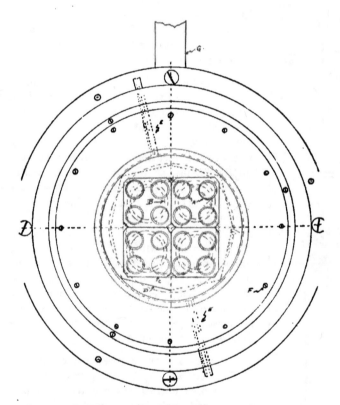

FIGURE 142—Front view of Egner-Holmström high current transmitter.

for the unusually large amounts of energy necessary in radio telephony are to replace the felt packing of the microphone chamber by asbestos packing or packing of some other uninflammable material, and to perforate the metal case so as to permit air cooling.

One of the most remarkable and effective of high current microphones is that devised by Messrs. C. Egner and J. G. Holmström of Stockholm,

Sweden. The inventors state that a normal microphone which, for a current of a few milliamperes has a resistance of say 200 ohms, at a current of 1 ampere has a resistance of only 5 to 8 ohms. The microphone is shown in plan in Figure 142 and in actual appearance in Figure

FIGURE 143—Egner-Holmström high current microphone.

143. Corresponding parts are indicated by the same lettering. The whole device is provided with oil (or other fluid) cooling, by the attachment of the cooling reservoir H (of part A of Figure 143) to the back of the microphone chamber. Through this cooling chamber run the supporting and connecting rods from each of the microphones A. It will be noticed from Figure 142 and part B of Figure 143 that there are 16 of these microphones, which can be connected together in various ways as indicated below. The rods which run from the microphones through the cooling reservoir terminate on the connecting board J (part C of Figure 143). The cooling fluid is arranged to circulate in a fashion similar to the "thermo-syphon" system sometimes used for gas engines, and heat is radiated from the flanges of the cooling chamber H. The cooling fluid must be an insulator.

The individual microphones are connected together permanently in 8 sets of 2 each, the 2 being always adjacent on the same row. The back of the microphone chamber is, in each case, a copper plate covered with thin carbon, and is fixed. From the copper plates pass the rods to the rear connecting board, previously mentioned. The vibrating electrodes C (of Figure 142 and part B of Figure 143,) are 4 in number, each taking care of 4 of the microphones back of it. These microphones are insulated from each other by being supported on cylinders of glass B

(part *B* of Figure 143) which cylinders are in turn attached to the main vibrating diaphragm. The reason for the use of glass or a similar poor conductor of heat is that it is desired to prevent overheating of the main diaphragm since this has been found to lead to speech distortion. The individual microphone chambers are made up of rings of asbestos or a similar heat-resistant material, pressed by spiral springs against the electrodes *C* so as to close the microphone chambers.

The main diaphragm is a thin sheet (0.2 mm. or 0.008 inch) of aluminum or magnalium which is stretched as tightly as possible. The stretching is accomplished by tightening up, one after another, the screws *F* (Figure 142 and part *A* of Figure 143). Since the 4 vibrating electrodes *C* are attached rigidly to the central portion of the main diaphragm, they will vibrate in the same phase and amplitude. It is this fact which renders it possible to secure a stable arrangement of microphones in parallel in the Egner-Holmström transmitter.

In order to increase the internal resistance and resistance variations of the transmitter, hydrogen or some hydrogen-containing gas is passed through the microphone chamber by means of the inlet and outlet pipes *E*. Normally the gas supply required is practically nil after the air originally present in the microphone chamber has been displaced.

The various ways in which the individual microphones can be connected are shown in Figure 144. These are as follows:

(a) 8 microphones in parallel, each of 2 in series. Proper applied voltage—10-15 volts. Proper current—up to 20 amperes.

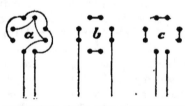

FIGURE 144—Connection arrangements of Egner-Holmström high current microphone.

(b) 4 microphones in parallel, each of 4 in series. Voltage—20-30 volts. Current—up to 10 amperes.

(c) 2 microphones in parallel, each of 8 in series. Voltage—40-60 volts. Current—up to 5 amperes.

It will be seen that the microphones can handle up to 200 to 300 watts (12 to 18 watts per individual microphone). The usual current (corresponding to case (b) above) is 10 amperes, but the makers of the transmitter, the Aktiebolaget Monofon of Stockholm, are prepared to build the transmitters to carry up to 16 amperes under these conditions There are about 0.3 cubic centimeter (0.018 cubic inch) of carbon granules in each individual microphone.

Messrs. Egner and Holmström tried out their transmitter in connection with Professor Poulsen's apparatus shown in Figure 13. On June 29th and 30th, 1909, as previously stated, clear communication was achieved using this transmitter with 6 amperes in the antenna between Lyngby and Esbjerg, a distance of 170 miles (270 km.).

Another form of microphone transmitter of considerable interest was used by Mr. R. Goldschmidt of Laeken (near Brussels) in conjunction with the apparatus shown in Figure 67. The device in question is

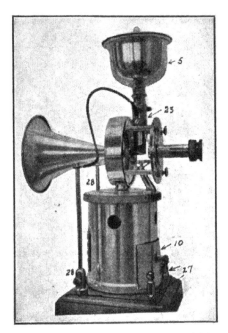

FIGURE 145—Scheidt-Boon Marzi high current microphone transmitter (relay type).

the invention of Mr. J. B. Marzi of Cornigliano (Liguria, Italy). The basis thereof is an attempt to prevent burning of the carbon grains when heavy currents are used by the expedient of using a *moving stream* of carbon grains. Very finely powdered carbon will flow in practically the same manner as a liquid stream, and a portion of the carbon stream, passing between two electrodes, is used in this case as the microphone. The actual apparatus is shown in Figure 145, and the cross sections of several forms thereof and the mode of connection are given by Figure 146.* Referring to parts I, II, and III of the latter Figure, a reservoir

---

* Figures 145 and 146 are reproduced by permission from the French journal "T.S.F." and based on material from Mr. Scheidt-Boon of Brussels (1914).

5 is filled with finely powdered carbon and from this reservoir a fine
stream of carbon flows through the hollow pipe 6 till it is compelled to
pass between the platinum surfaces 9. These may be portions of concentric spheres (as in part I), or an obliquely cut cylinder and a plane

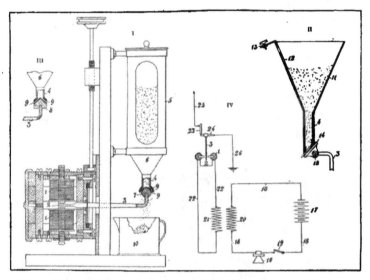

FIGURE 146—Details of Marzi high current microphones and relays.

surface (as in part II), or portions of two coaxial cones (as in part III).
In any case the carbon streams between these surfaces, which are the
terminal electrodes of the high current microphone. The upper one of
these surfaces is usually fixed whereas the lower one is movable, either
by the voice directly or, as shown in part I, by means of an armature 2
controlled by the electromagnets 1, 1. The current for these electromagnets is derived from the circuit of an ordinary telephone transmitter, or
from a telephone line. It is this feature which makes the device a relay.
In Figure 145 the terminals 27 are those of the electromagnets 1, 1 and
the terminals 28 are the heavy current microphone terminals.

After passing the surfaces 9, 9, the carbon stream flows into the cup
10. At regular intervals, the contents of this cup should be emptied
back into reservoir 5. The circuit diagram is indicated clearly in part
IV of Figure 146. As will be seen, the ordinary microphone circuit is
coupled through the induction coil, 20, 21 to the circuit containing the
electromagnets 1, 1 of the relay. The high current transmitter is shown
placed directly in the antenna, though it can equally well be employed

in any of the ways shown under ordinary "Microphone Transmitter
Control." (page 133.) The weight of the entire apparatus is only about
9 pounds (4 kg.) and the height thereof 18 inches (45 cm.). As previously
stated, this transmitter, carrying 3 amperes, permitted communication
from Laeken to Paris, a distance of 200 miles (320 km.).

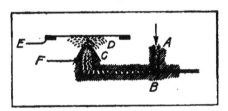

FIGURE 147—Essential parts of Chambers
liquid microphone.

Another method of attacking the problem of high current microphones has been the attempt to use conducting liquid jets of one type
or another. Figure 147 shows the essential parts of a simple microphone
of this sort devised by Mr. F. J. Chambers in 1910. At A a stream of
electrolyte under a head of about 3 feet (1 m.) flows past the needle
valve B. Here the flow is adjusted to a suitable amount. The liquid
then passes through the conducting nozzle C, which is connected to F,
one of the terminals of the microphone. After leaving the nozzle, the
liquid stream impinges on the diaphragm D which is vibrated by the
voice. The diaphragm is suitably connected to E, the other terminal
of the microphone. It will be seen that the up-and-down motion of the
diaphragm will alter the length and cross sectional area of the jet and
consequently its resistance. It is found that such a microphone, because
of the mechanical damping of the diaphragm by the jet, gives clear
articulation without rasping noises. The distance of the diaphragm
from the nozzle is adjustable. The capacity of such a microphone is
limited simply by the necessity of preventing the current-carrying electrolyte from boiling. In practice, Mr. Chambers found that about 400
watts could be handled by such a microphone.

Another type of liquid microphone, somewhat similar to that of Mr.
Chambers, has been devised by Professor Giuseppi Vanni of Rome.
The apparatus is shown in Figure 148. A centrifugal pump R, made
entirely of acid-resistant materials and operated by a small motor,
forces a jet of dilute acid out of the ebonite nozzle T. The jet then falls
on the inclined surface A, is deflected to the oppositely inclined surface
B, is again deflected and then passes back to the pump to resume its
circulation. The pump pressure corresponds to 12 or 15 feet (3 or 4 m.)

of water column.   The terminals of the microphone, H and P, are connected mechanically to the electrodes A and B. B is fixed but A is vibrated back and forth in an oblique direction practically perpendicular to the deflected jet. Z is the mouthpiece of the microphone,

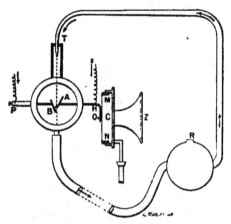

FIGURE 148—Vanni's liquid microphone.

the diaphragm being connected at O to the mechanical control of A. The motions of A not only change the cross section of the jet from a cylinder to a flattened sheet, but also obstruct the stream more or less by the greater or less immersion therein of A.   The electrode A therefore acts as a sort of shutter.

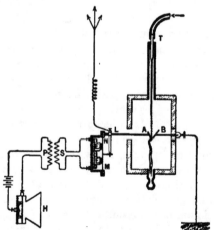

FIGURE 149—Vanni's liquid microphone
relay.

The Vanni microphone has also been arranged as a relay in the fashion illustrated in Figure 149. The usual transmitter H supplies fluctuating currents to the electromagnets E, which in turn control the liquid microphone by means of the iron diaphragm NM. This device was used in the experiments of Professor Vanni previously described, where it successfully controlled 1 kilowatt, permitting radio telephony 625 miles (1,000 km.). These experiments were described on page 71.

Another principle which can be applied in liquid microphones is that of the instability of liquid jets, as first discovered by Chichester Bell in 1886. The phenomenon, which is based on the surface tension of the liquid, is illustrated in Figure 150, part A. This shows a jet of liquid escaping from a small tube, T. The orifice of the tube is supposed to be smooth and circular. The jet will proceed as a cylinder for a certain distance, and then a slight constriction will occur at the point A. Directly below A the stream will bulge, and then constrict still more below the bulge. At B the stream will break up into drops which, as they fall, will vibrate from oblate to prolate ellipsoids, passing through the spherical shape. We are not, however, concerned with the stream after it has broken up, but rather with its cross-section at the bulge just above the

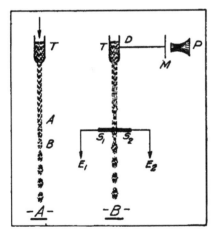

FIGURE 150—Essential parts of Majorana's liquid microphone.

breaking-up point. For it is found, by experiment, that the transmission of the least mechanical disturbance to the falling jet will move the breaking-up point up the stream toward the orifice, and the motion will be quite considerable even for very slight mechanical disturbances.

These facts have been utilized by Professor Q. Majorana of Rome in his hydraulic microphone, devised in 1906. Its essential parts are shown in part B of Figure 150. The tube T of glass or other insulator has a portion of its wall at D replaced by an elastic diaphragm which is attached to the larger voice-actuated diaphragm M by mechanical means. Placed in the jet just above the breaking-up point are the two electrodes S_1 and S_2 which form the terminals of the microphone. It is clear that the variations of cross sections of the jet at S_1S_2 will cause the necessary resistance variations. One unfortunate drawback with this form of liquid microphone is the excessive length of the jet (5 to 15 feet, or 2 to 5 m.) and its very great sensitiveness to slight shocks. As previously described, Professor Majorana succeeded in telephoning 270 miles (420 km.) with such a microphone control. It has been stated that the device can control 10 amperes at a terminal potential difference of 50 volts, corresponding therefore to 500 watts.

CHAPTER VII.

(g) VACUUM TUBE CONTROL SYSTEMS.

As has been previously described in considerable detail, a ready means of generating moderate, and even high outputs at sustained radio frequencies is by the use of the various types of hot cathode vacuum tubes. These tubes depend for their operation as oscillators on the potential of small conducting members such as the grid in audions, oscillions, and pliotrons. Since the amount of energy required to change the potential of small-capacity conducting members is itself minute, it would seem *a priori* that one of the most ready means of modulating the output of such oscillators would be by altering the potential of the member in question in accordance with the voice vibrations. As a matter of fact, the proper control of the oscillations generated in such a tube is not a perfectly simple matter, for reasons which will appear.

There are at least two available methods of controlling the output of vacuum tube oscillators, and instances of each of these in practice will be described. The first of these is by variation of the grid potential, the assumption being that as the grid potential becomes increasingly negative, the current through the tube (and therefore the available radio frequency output) continuously and proportionately diminishes. Difficulties of stability of operation, however, arise and the conclusion must be somewhat modified. The second of these methods is by varying the plate potential, the assumption in this case being that as the plate potential becomes increasingly positive, the current through the tube (and therefore the available radio frequency output) continuously and proportionately increases. This conclusion also requires some modification because of temperature and space charge limitation of plate current and because of the limits of available energy which must thus be introduced into the plate circuit. (See under ''Sustained Wave Generators,'' part (c), ''Vacuum Tube Oscillators;'' and specifically the descriptions given of Figures 71, 72, and 73, page 77.)

FIGURE 151—Telefunken Company-Meissner
radiophone transmitter, 1913.

There are a number of differences between the operation of the two systems of modulation mentioned, such as the relation between the modulated radio frequency energy and the necessary controlling audio frequency energy; but these differences will best be brought out in considering the actual systems in use.

Dr. Alexander Meissner of the Telefunken Company, working with the tube shown in Figure 83, and the circuit shown in Figure 81 for producing the oscillations, succeeded in carrying out some interesting experiments in radiophone transmission. He states that using a plate circuit voltage of 440, it was possible to obtain a radio frequency output of 12 watts in the antenna. This corresponded to an antenna current of 1.3 amperes with an antenna resistance of 7 ohms at a wave-length of 600 meters. No statement was made as to the mode of control, though for such small powers it is probable that a heavy current microphone would suffice if placed directly in the antenna or in a suitably associated circuit as indicated under the description of Figure 130, page 133.

The radiophone equipment used by Dr. Meissner in June, 1913, for transmission between Berlin and Nauen, a distance of 23 miles (36 km.), is shown in Figure 151. The von Lieben-Reisz bulb is mounted at the rear of the apparatus box.

While these experiments were significant, it must be noted that Mr. H. J. Round states that when the Lieben-Reisz tubes were used at such outputs, they lasted only 10 minutes because of disintegration of the filament by the positive ionic bombardment! This would naturally render their use under such conditions impracticable commercially.

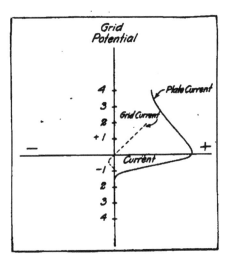

FIGURE 152—Plate-to-filament and grid-to-
filament characteristics of three electrode hot cathode tube containing gas.

We consider next the radiophone experiments carried out by Mr. Round of the Marconi Company. To begin with, we shall give the grid potential-plate current curves found for his tubes by Mr. Round. One of these is shown in Figure 152. It should be carefully compared with that shown in Figure 76 for the case of pure electron discharge tubes. Mr. Round's description of Figure 152 (with some added comments and slight alterations) will be given: "Suppose the plate to be made so positive that the whole tube would be glowing (i. e., filled with blue glow of the usual ionised gas discharge) except for the presence of the grid. Then, starting with the grid strongly negative, notwithstanding the plate being highly positive, the electrons cannot get through the grid because the grid is nearest to them. At a very small negative value of the grid potential, a few electrons can get through the grid and will fall to the plate and the number that will get through will rapidly increase until the grid is at zero potential; the current to the plate then having the value it would if the grid were absent. Afterwards, as the grid becomes positive, the current will decrease because the grid will absorb some electrons.''

The detailed wiring of a Marconi Company radiophone transmitter (the receiving set of which is described on page 216) is given in Figure 153. It will be seen that oscillations are produced by

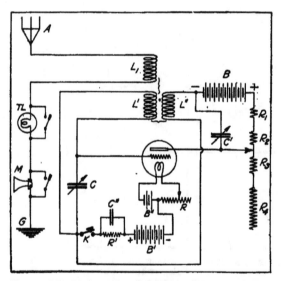

FIGURE 153—Marconi Company radiophone transmitter.

coupling the grid circuit $L'C$ with the plate circuit $L''C'$ by means of the inductive coupling $L' L''$. The grid circuit also contains the 30 volt battery B' and the 3,500 ohm resistance R', which latter is shunted by a suitable by-pass condenser permitting the transfer of radio frequency currents but preventing excessive direct grid current. Similarly, the plate circuit also contains the resistance R_1, R_2, R_3, each of which is 500 ohms, and the resistance R_4 of 10,000 ohms. These prevent excessive plate current, "blue glow," and tube breakdown. In series with these is the plate battery B of 500 volts. The aggregate of resistances and battery is shunted by the capacity C' of the plate oscillating circuit. The radio frequency energy thus produced is transferred to the antenna circuit at L_1 by an inductive coupling. The presence of oscillations in the antenna is indicated by glowing of the test lamp TL which can be short-circuited when not in use. The microphone M is directly inserted in the antenna circuit, and can also be short-circuited for purposes of tuning. The battery B'' used for lighting the filament is an ordinary 80 ampere-hour storage battery. The battery B' for providing 500 volts consists of four cases of dry cells. These were found suitable for the needs of the occasion since only 10 to 20 milliamperes (0.010 to 0.020 ampere) were required. Thus the input is from 5 to 10 watts. The set is arranged so

FIGURE 154—Marconi Company radiophone set.

that it can also be used for telegraphy by manipulating the key K in the grid circuit. The change-over switch from sending to receiving is simple and is so arranged that it can be controlled from a distance thus permitting handling the set from any part of a ship, e. g., the chart room. Needless to say, the duplicate transmitter (microphone) and receiver could also be placed there. The set delivers 0.6 amperes in the antenna, and is guaranteed for communication over 30 miles (50 km.) with ship antennas 100 feet (30 meters) high and 200 feet (60 meters) apart. The set can, however, be pushed to give 1 ampere in the antenna with an estimated sea range of 100 miles (160 km.). As a matter of fact, communication was established with such a set between Aldene, New Jersey, at the station of the Marconi Company and a station in Philadelphia,

Pennsylvania, a distance of 65 miles overland (105 km.). The Aldene antenna was supported on two 200-foot (60 m.) towers 450 feet (145 m.) apart. The actual appearance of the set is given by Figure 154. The large generating valve is shown at V between the vertical supports. To its right is placed the small receiving valve.

It is stated by Mr. Round that the telegraphic range of these sets is twice the radiophonic range. The tuning is found to be unusually sharp, in fact, almost uncomfortably so. It was also found somewhat difficult to start these tubes rapidly in cold weather. Just before the war, work was proceeding with such equipment in the direction of a selective call system, but this had to be suspended.

Using tubes of the sort described, Mr. Round succeeded in getting 3 amperes in the antenna, which would probably correspond to about 50

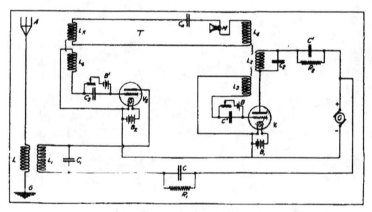

FIGURE 155—Marconi Company-Round radiophone transmitter of 1914.

watts output. The input was about 0.100 ampere at 2,000 volts or 200 watts, thus giving an efficiency of about 25 per cent. The efficiency here referred to is the so-called "electron efficiency"; that is, it does not include in the input the energy required for lighting the filament, but considers only the plate circuit input and the radio frequency output. Mr. Round considers 2,000 volts to be excessively high for tubes of this sort, and states that experiments are being conducted whereby it is hoped to make the tubes available for use at lower voltages. The serious objection to lower voltages is that the high supply currents then required for appreciable outputs produce very rapid filament disintegration if gas be present.

Another form of the Marconi Company's radiophone transmitter is shown in Figure 155. Here a master oscillator, V_1, is used, the output of which passes through an intermediate circuit T to the grid circuit of an amplifier, V_2. The output of amplifier V_2 is transferred to the antenna through an inductive coupling. The modulation control in this system is accomplished by placing a microphone in the intermediate circuit thus varying the radio frequency voltage impressed on the amplifier grid. This system of master oscillator and amplifier is of considerable interest and the illustration shows one of the earliest forms thereof. The details of the master oscillator, V_1, are seen to be those of Figure 153. The amplifier V_2 is very similar except that its grid and plate circuits are not coupled. It will be noted that both the master oscillator and the amplifier are fed from the same plate generator G. Here we have a case where the microphone does not have to handle the whole of the antenna energy, and indeed the amount handled by the microphone M is roughly the antenna energy divided by the amplification produced in V_2. A modification of this system omits the amplifier but uses the microphone as part of the coupling between L_2 and L_3 in the master oscillator, thus suitably varying its output.

A recent type of Marconi Company sustained wave bulb transmitter is shown in Figure 156. As will be seen, it includes two bulbs, each enclosed in a ground-glass front compartment. The three instruments at the top of the case indicate the plate current, antenna current, and filament current. Means for regulating the filament currents, for tuning the various circuits, and for varying the regenerative coupling between the grid and plate circuits are provided.

We consider next the oscillion radiophone transmitters manufactured by the de Forest Radio Telephone and Telegraph Company. One type of these is illustrated diagrammatically in Figure 157, and this arrangement of circuits is due to Mr. C. V. Logwood. It will be seen that the direct current generator G (usually of 1,200 to 1,500 volts) is connected in series with the iron core choke coil L' and shunted by the condenser C', the purpose of these being to cut down the "commutator ripple" thus giving a more nearly constant e. m. f. in the plate circuit. Failure to observe this precaution leads to a loud and objectionable hum corresponding to the frequency with which the commutator segments pass under the generator brushes. The oscillating circuit used seems to be of the capacitive coupling type or ultraudion type according to the method of classification. It is clear that the antenna capacity is used as a portion of the oscillating circuit and that the antenna resistance is used to absorb the output of the bulb. Modulation is accomplished by impressing audio frequency potential variations, produced by the voice, on the grid. The microphone M causes varying currents in the circuit of the transformer primary P, whence the potential variations are produced by the secondary S in the filament-to-grid circuit. The resistance R_1 in series with S serves to keep the grid strongly negative because of the difficulty experienced

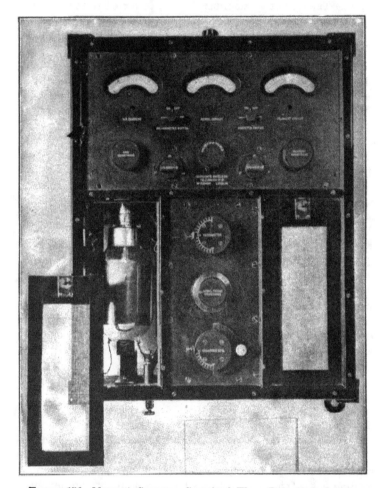

FIGURE 156—Marconi Company Sustained Wave Bulb Transmitter.

by any negative charges on the grid in leaking off to the filament. By varying R_1, the grid potential can be varied. For telegraphy, the key K is used. It merely opens the grid leak circuit, whereupon the grid immediately becomes so negative as to choke off all plate current and thus stop the oscillations entirely. Closing the key permits the excess negative charge to leak off the grid and the oscillations start again.

There is a marked tendency to increase the dimensions and available output of the tubes employed, and this is well illustrated in Figure 158. The left hand tube is of approximately the dimensions of the usual ampli- fier or ''repeater'' bulbs used by the Western Electric Company in trans- continental wire telephony. This company is operating under exclusive patent licenses granted by the de Forest Company. The right hand bulb is one of the latest 0.25 kilowatt input oscillions. A 3 or 4-inch (7.5 or 10 cm.) ''laboratory oscillion'' is shown mounted on its panel in Figure 159.

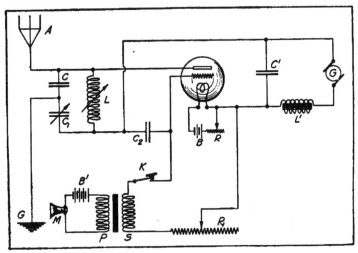

FIGURE 157—de Forest Company-Logwood radiophone transmitter.

Such a device can produce conveniently a number of watts of radio fre- quency energy of constant amplitude.

A whole series of radiophone transmitters have been put on the mar- ket by the de Forest Company some of which are here illustrated. A low power set using what is practically one of the tubular receiving bulbs is seen in Figure 160. A larger type of transmitter and receiver, together with the requisite motor generator set, appears in Figure 161. This set is stated to have a telegraphic range over water of 40 miles (64 km.) using masts 200 feet (60 m.) high and an antenna span of at least 250 feet (80 m.). The generator of the motor generator set in this case is for 1,000 volts and 100 watts output. Though detailed wiring diagrams of the arrangement shown in Figure 161 are not available, it is of some interest. It shows a complete de Forest radiophone trans- mitter and receiver. At the left is shown the bulb-mounting panel. Dr. de Forest has given the name of ''oscillion'' to the bulb shown in the figure. This bulb has a tungsten ''W'' filament, a grid of tungsten wire wound on a glass support, and two nickel plates. As seen from the figure, the bulb is air-cooled by means of the small fan placed underneath it. The two instruments mounted on top of the panel are respectively indicators of the filament amperage and plate circuit current of the oscillion. The switch at the left hand side turns the plate current of the tube on and off. The filament current control-rheostat handle is shown in the lower right hand corner of this panel. In the middle box are mounted the various portions of the oscillating circuits and microphone control apparatus. The microphone transmitter is visible on the front. The equipment to the right of the figure is a fairly normal audion receiving set. A more recent set of this type is shown in Figure 163. The small ammeter at the top left indicates the filament current of the bulb, which requires somewhat careful setting for full output. The right hand top instrument is the antenna ammeter. A convenient form of protected change-over switch from sending to receiving is mounted on the back of the panel, the handle projecting just to the right of the microphone arm. The bulb is also mounted back of the panel, and can be partly viewed through a slit under the microphone arm. The variable condenser to the left of the arm is condenser C_1 of Figure 157. A filament rheostat and binding posts for the filament battery, antenna and ground connections, etc., completes the installation except for a short-circuiting bar between two binding posts. This latter may be removed and replaced by the Morse key, then permitting telegraphy.

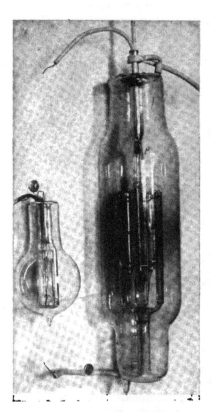

FIGURE 158—de Forest 0.25 K.W. oscillion (as compared with type of bulb used in long distance wire telephony.)

An extremely interesting aeroplane radiophone transmitter is shown complete in Figure 164. The generator is driven by the air propeller with suitable speed control devices, and is enclosed in the "stream line"

FIGURE 160—de Forest low power radiophone transmitter (Type PJ).

FIGURE 159—de Forest laboratory oscillion transmitter.

casing, the terminal leads being brought out of the rear end. The oscillion is mounted in a protective wire mesh casing and is suspended in such fashion as to be reasonably safe from breakage. The three top instruments are for antenna current, plate circuit current, and filament current. The Morse key is shown at the bottom of the figure together

FIGURE 161—de Forest "oscillion" radiophone transmitter (and receiving set).

with the microphone. The latter is so arranged as to fit closely to the lips of the user and thus avoid picking up the extremely loud noise of the engine exhaust.

A more elaborate type of radiophone transmitter using three oscillion bulbs is shown in Figure 165. It includes a "modulator" or master oscillator bulb and two "radio" or amplifier bulbs. These are mounted back of the panel, and can be viewed through the three slits. The four instruments at the top of the board (starting at the left) are respectively

FIGURE 162—de Forest oscillion radiophone and radio telegraph transmitter and receiver (Type OJ3).

for the "modulator" current, plate circuit current, filament circuit current, and antenna current. The inductance and variable condenser of the master oscillator circuit are mounted directly below the corresponding ammeter at the left. An antenna loading inductance and a control switch for changing from receiving to transmitting are mounted to the right of

FIGURE 163—de Forest 0.25 K.W. oscillion radiophone transmitter.

the microphone transmitter. Under the slit of each bulb are its filament and plate circuit switches, and at the bottom of the board are the three filament rheostats. As before, two binding posts are provided at the bottom of the board for the insertion of a Morse key if radio telegraphy is desired. A set of this type is supplied with 1,500 volts for the plate circuit and an input of about 1.5 kilowatt. The telephone range over water is stated to be 400 miles (640 km.) and the corresponding tele-

FIGURE 164—de Forest 0.25 K.W. oscillion radiophone and radio telegraph transmitter.

graphic range 600 miles (1,000 km.). As before, towers 200 feet (60 m.) high and 250 feet (80 m.) apart are presupposed.

Arrangements were made by the de Forest Company with a phonograph company whereby almost every night records made by this latter company were played into the radiophone transmitter and thus rendered audible to a wide circle of listeners. One or two oscillions are used in the transmitter, each with a stated output of 0.25 k.w. The wave length used has been 850 m. This service was given about five nights per week beginning October, 1916. The music has been heard a number of times as far away as Buffalo, New York, a distance of 306 miles (490 km.) and even at an extreme range at Mansfield, Ohio, a distance of 465 miles (750 km.).   One interesting result of this work has been a "radio dance" given one evening at Morristown, New Jersey, a distance of 30 miles (50 km.) from the de Forest station. Music was transmitted from the latter station and received at Morristown on a receiving set with a three-step audion amplifier.   The resulting "signals" were sufficiently loud to permit the dance to be conducted.   Another novel field for radio telephony, which Dr. de Forest believes presents great promise, is that of news distribution in rural districts.   There is no doubt that the dissemination of

FIGURE 165—de Forest 0.25 K.W. 3-oscillion radiophone transmitter.

information and various types of entertainment in districts which would otherwise be isolated is a most valuable possibility for radio telephony.

As is well known, the Western Electric Company has been carrying on extensive research work in radio telephony for some time past. (Some of the types of tubes described in the patents of that Company are similar to those shown in Figure 95.   Generally speaking, platinum filaments coated with metallic oxids are there indicated.)   A method of modulation of the output of such an oscillator has been developed by Mr. E. H. Colpitts.   It is depicted in Figure 166.   As will be seen, the plate oscillating circuit $C_1L_1L_2C_2$ is coupled inductively to the grid circuit CL at L_1. It is also coupled inductively to the output circuit $L''C''$ at L_2.   A second grid circuit is also provided consisting of the secondary S of the audio frequency transformer (the primary of which contains the microphone and battery B), and the battery B_1 for maintaining the grid at a negative potential. This system of modulation has the advantage of simplicity. On the other hand, it may easily become an unstable control system. The reason for this is the following: In any oscillating tube, the amplitude of the plate circuit oscillations increases until the losses in the tube, and in the external or output circuits which it feeds, utilise the entire available energy. The amplitude then remains constant. It is evident that if we make the grid potential extremely negative, so that the plate circuit oscillations cannot build up to this stable value just mentioned, the oscilla-

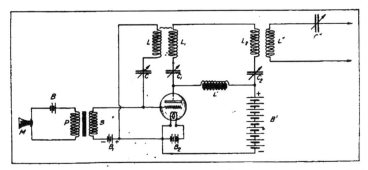

Figure 166—Western Electric Company-Colpitts modulation system, 1914.

tions will simply cease entirely. Just above this extremely negative grid potential, there is a narrow range of grid voltages for which the plate circuit output depends on the grid potential, though only as a transient phenomenon. A static characteristic of such a relation between grid potential and oscillating current in the plate circuit is not obtainable because the effects do not persist. The oscillating current tends to rise either to its full and stable amplitude or to cease altogether. For audio frequency variations of moderate magnitude and sufficient rapidity of the grid potential the system is sometimes workable though always with the danger just mentioned for low tones or for extremely loud sounds.

A second system due to Mr. Heising* of the same company is free from the objections mentioned in that the tube is used as an *amplifier* and not as an oscillator. The method in question is shown in Figure 167. The radio frequency source A impresses, through the transformer P_1S_1, corresponding radio frequency potential variations on the grid G_1 of the tube. There will, therefore, be produced in the output plate circuit of this tube radio frequency current variations. Hence there is an available

*Patent 1,199,180.

output in the inductance L_2. The tube has a second grid, G_2, and, as will be readily seen, there are impressed on G_2 potential variations corresponding to the speech amplitudes, these variations being produced in the customary way by a microphone circuit and a suitable transformer. The source A may naturally be a vacuum tube oscillator. Each grid is maintained at a suitable negative potential by the battery B_1 or B_2.

A series of long distance radiophone experiments were carried on by the Western Electric Company from the United States Naval Radio Station at Arlington, Virginia. This station has an antenna 600 feet (180 meters) high. Speech was transmitted by night from Arlington to the Eiffel Tower, Paris (a distance of 3,900 miles or 6,200 km. almost entirely over water), from Arlington to Mare Island, California (a distance of 2,400 miles or 3,800 km. overland), and from Arlington to Hawaii (a

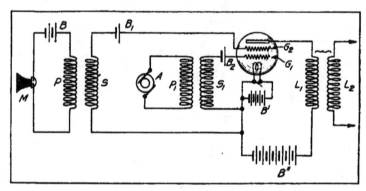

FIGURE 167—Western Electric Company-Heising modulation control system, 1915.

distance of 5,100 miles or 8,300 km., about half over water). While the transmission could be achieved only under exceptional conditions and was in no sense commercial, it is of marked interest in indicating how great a distance can be bridged by even a very moderate amount of power under favorable circumstances. One is reminded of the feat of Sayville, Long Island, in communicating with Nauen, Germany, a distance of 4,200 miles (6,700 km.) with only 6 kilowatts in the antenna.

The apparatus used at Arlington was constituted as follows: A small bulb (3 inches or 7.5 cm. in diameter) was used as a master oscillator. The filament was heated from storage batteries as usual, and the plate circuit was fed from 125 volts in dry batteries. The master oscillator had a fairly fine grid. Its output circuit was coupled loosely to the grid circuit of a 7-inch (17.7 cm.) "modulator" bulb with a coarser grid. Comprised in this grid circuit were a 150 volt battery, to give the grid the requisite negative potential, and the secondary of a 150-to-1 air core transformer in the primary of which was a button-type microphone and its supply battery. In this way, the voice potential variations were impressed on the modulator grid as well as the radio frequency variations. The plate circuit of the modulator was tuned, and included a 450-volt direct current generator.

The output circuit of the modulator supplied speech-modulated, radio frequency, potential variations to the fairly coarse grids of 7-inch (17.7 cm.) bulbs all connected in parallel. Their tuned output circuit in

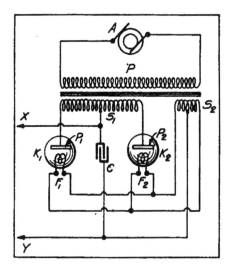

FIGURE 168—General Electric Company-
White method of producing practically constant potential from A. C.

turn fed the coarse grids of from 300 to over 500 "power" bulbs in parallel. As before, these grids were kept at a constant negative potential of —150 relative to their filaments. The plate circuit of the "power" bulbs was fed from a large 600-volt, direct current generator which was normally used for the Poulsen arc at Arlington. A few turns of heavy copper band in this last plate circuit were inductively coupled to the tuned antenna. About 60 amperes at 6,000 meters wave-length were normally produced in the antenna, this corresponding to something over 9 kilowatts. The efficiency of the set was about 20 per cent. In running the set, fairly frequent bulb renewals were required, thus rendering a high upkeep cost of operation inevitable (according to one statement, $10,000 per month).

The apparatus used was mounted on a series of panels. The lower section of each panel had the necessary switches for controlling the filament and plate circuits of that section. The upper portion of each panel was in two halves. On each half were mounted 25 of the 7-inch (17.7 cm.) "power" bulbs, all cooled by air brought in ducts from a powerful blower. The cooling ducts were at the rear of the panel. All the bulbs on each panel portion were in parallel. Each bulb was provided with "Ediswan" socket base so as to be readily replaceable, i. e., all terminals were brought out through this base. The control and modulator bulbs were mounted on separate small panels.

We consider next a number of radiophone pliotron transmitters designed by the Research Laboratory and especially Mr. William C. White of the General Electric Company. The mode of producing reasonably constant sources of high potential (from alternating current supply) will be first considered.

The method referred to is illustrated in Figure 168. The alternator *A* sends current through the primary *P* of a transformer. This trans-

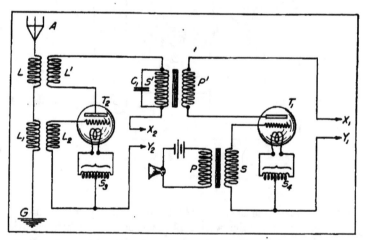

FIGURE 169—General Electric Company-White radiophone transmitter for alternating current supply.

former has two secondaries. Of these one, S_2, is arranged to light the filaments F_1 and F_2 of two kenotron rectifiers. There are comparatively few turns in the secondary S_2 because the filament voltage is low. A second secondary S_1 is of many turns so as to furnish a high voltage to the plates P_1 and P_2 of the kenotrons. It will be noted that there is a central tap of the filament-feeding secondary S_2 the purpose of which is explained in connection with the description of Figure 74. It prevents injuriously excessive addition of the filament-heating and thermionic currents in either end of the filament. The middle point of the secondary S_1 is connected to one side of a large high voltage condenser C (e. g., of several microfarads), the other side of which condenser is connected to the middle point tap of the filament-heating secondary S_2. It will be seen that the condenser will be charged during one-half of the cycle by the left hand half of S_1 in series with kenotron K_1 and during the other half of the cycle by the right hand half of S_1 and the right hand kenotron K_2. If the current drawn from the charged condenser is comparatively small (which will be the case if the condenser is very large and a small current at high voltage is drawn therefrom), the potential difference at its terminals will remain appreciably constant. Experience shows indeed that this is the case, and it has proven possible to get so nearly constant a potential from an alternating current supply in this way that, when used in the plate circuit of a normal pliotron oscillator, the usual a. c. hum has been practically absent. The output is drawn from the condenser terminals, X, Y.

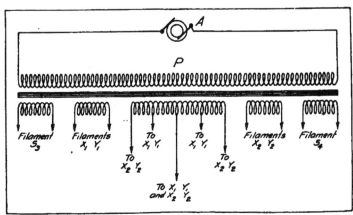

FIGURE 170—General Electric Company-White multiple transformer for feeding plate rectifier and filaments.

Two of the earlier types of radiophone transmitters based on this principle will be next described, the description being due to Dr. Irving Langmuir of the General Electric Company.*

"The first outfit has a capacity of about 20 watts in the antenna, the source of power being the local city supply, which is 118-volt, 60-cycle current. This is connected with the primary of a small transformer having two secondary windings. One of the secondaries is designed to

* "Proceedings of the Institute of Radio Engineers," Volume 3, number 3, September, 1915.

give about 5 volts and furnishes the currents used for heating the filaments of the kenotrons and pliotrons. The other secondary of the transformer is wound to furnish a potential of about 800 volts. This is rectified by means of a kenotron, and serves to charge a condenser of about 6 microfarads. In this way, a source of high voltage, direct current is obtained in a very simple manner. The plate of the pliotron oscillator is then connected to one of the terminals of the condenser, while the filament is connected to the other. The plate of the second pliotron is connected to the grid of the first, while the grid of the second is coupled by means of a second small transformer to the microphone circuit. With this small outfit, both pliotrons may be relatively small. . . .

"In the second outfit, which is suitable for use up to 500 watts or

FIGURE 171—General Electric Company-White radiophone transmitter for direct current supply.

more, the high voltage direct current is obtained from a small 2,000-cycle generator. The current from this is transformed up to about 5,000 volts, rectified by kenotrons, and smoothed out by means of condensers. By the use of 2,000-cycle alternating current instead of 60-cycle, it is possible to store up large quantities of energy at a given voltage and with a permissible fluctuation of voltage, and thus obtain as much as a kilowatt or more of power in the form of direct current with condensers of moderate size. This high voltage direct current is used, as before, to operate a pliotron oscillator, the output of which is controlled by means of a small pliotron connected to the telephone transmitter. . . . '' Wire-line-to-radio telephone transfer has been accomplished with such sets.

Another form of radiophone transmitter of the General Electric Company, described in Mr. W. C. White's patent 1,195,632, is shown in Figure 169. It will be seen that the grid of pliotron amplifier T_1 is connected to the filament through the secondary S of a transformer, the primary of which contains a microphone and battery. The plate circuit is fed at $X_1 Y_1$, by exactly the same form of device as shown at XY in Figure 168. For the sake of simplicity, this device is not here repeated in the diagram. The output of pliotron T_1 is fed into the plate circuit of pliotron T_2 through the audio frequency transformer $P'S'$. The secondary of this transformer is shunted by the condenser C_1 which acts as a practically perfect by-pass for the radio frequency currents in the plate circuit of T_2 without passing any appreciable quantity of audio frequency current from S'. It will be seen that the tube T_2 is an oscillator since its grid and plate circuits are coupled through the antenna circuit at $L\ L'$ and $L_1 L_2$. Obviously, the method of modulation control here shown is an extremely stable one. It consists in varying the plate potential of oscillator T_2 in accordance with the speech. This implies, however, the injection of considerable energy into the plate circuit of T_2 intermittently and hence the necessity for amplifier T_1.

For use with a radiophone outfit of this sort, a special transformer shown in Figure 170 may be used. This has the single primary P but a number of secondaries which supply the following circuits (starting from the left) : filaments of the oscillator T_2, filaments of the kenotrons which feed the amplifier T_1, plate circuits of the kenotrons feeding the amplifier T_1 and the oscillator T_2 (at different voltages, and the greater for the oscillator), filaments of the kenotrons feeding the oscillator T_2, and filaments of the amplifier T_1. Thus the entire set is started by closing one primary circuit, an obvious advantage.

A radiophone transmitter for direct connection to 125 volt direct current circuits is shown in Figure 171. The plug at the left of the set is merely inserted (with correct polarity) into a lamp socket and the change-over switch thrown to ''transmit'' in order to start everything in the set. It will be seen that the set is self-contained. The usual microphone transmitter, which can be a distance from the remainder of the set, is seen on the top of the box. Only direct current (obtained by bridging the microphone across a portion of a 125-volt potentiometer) passes through the microphone. At the top of the box at the left is mounted a small fixed condenser which is placed across the feeding line to reduce commutator ripple and to act as a radio frequency shunt in the plate circuit. Thus the 125-volt current feeds the plate circuit of the pliotron which is mounted inside the various coils. The filament is lit from the 125-volt circuit through an appropriate resistance. These various resistances and potentiometer are shown in the foreground at the bottom of the box. The two left hand coils are the grid circuit coupling to the antenna and the coils at the right the plate circuit coupling, a circuit

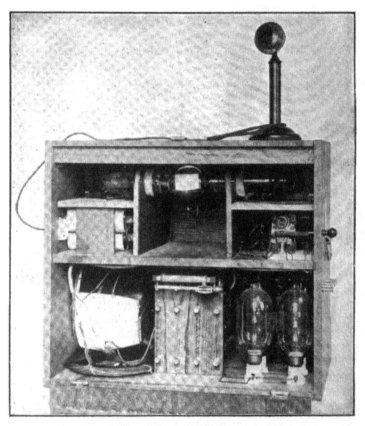

FIGURE 172—General Electric Company-White radiophone transmitter for alternating current supply.

somewhat like that in Figure 169 being used. The entire set weighs only 54 pounds (20 kg.) complete. Completely satisfactory operation over 10 miles (16 km.) is possible, and laboratory tests have given ranges up to 65 miles (105 km.).

A more powerful set for use with 60 cycle alternating current supply is shown in Figure 172. The wiring of this set is almost identical with that shown in Figures 168, 169, and 170. The two pliotrons are mounted at the top of the box. To the left, under them, are the microphone dry batteries. To the right, under them, are the "smoothing condensers" (two sets) for the high voltage supply in the plate circuits. To the bottom left are mounted the radio frequency coupling coils and to the right the four kenotron rectifiers. The panel in the middle carries various filament resistances, and back thereof are mounted the microphone transformer (*PS* of Figure 169) and the amplifier transformer (*P'S'* of the same figure). The entire set weighs 150 pounds (68 km.). The transmitting range for satisfactory service is 50 miles (80 km.).

We consider next the control systems suitable for use with the dynatron and pliodynatron tubes of the General Electric Company as developed by Dr. Albert W. Hull. A description of the dynatron (and

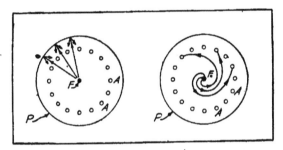

FIGURE 173—Effect of longitudinal magnetic field on electron paths in dynatron.

pliodynatron) together with their mode of operation is given in connection with Figures 96 through 101, page 100, and the reader is referred to this material as an introduction to the present discussion.

Figure 173 represents the cross section of a dynatron where *F* is the filament, *A* the wires, or solid portions, of the anode, and *P* the plate. The paths of a few electrons away from the filament and a diagrammatic representation of a few of the electrons leaving the plate by secondary emission are given for normal conditions in the left hand portion of the diagram. The effect on the electron paths of a longitudinal magnetic field (parallel to the filament) is shown in the right hand portion of the figure. It will be seen that the electrons now pursue spiral paths and strike the anode very obliquely, particularly if the magnetic field is very powerful and the electron velocity small. In consequence, comparatively few will get through the anode with a high velocity, and therefore the re-emission phenomena from the plate will be much diminished. The characteristics of the dynatron will be progressively altered, as indicated in Figure 174, whence the magnetic field is increased. The dotted curve, *A*, is the normal dynatron potential-current curve. On applying a moderate magnetic field the dashed curve, *B*, is obtained. This shows no current reversal since the secondary emission is already small. With a strong magnetic field, the characteristic becomes the full line curve, *C*, and shows very little of the usual dynatron effect. It is therefore pos-

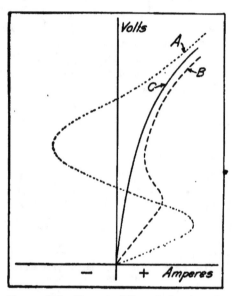

FIGURE 174—Characteristics of dynatron in
various magnetic fields.

sible to control the negative resistance (and hence the output) of a dynatron by the superposed magnetic field, and this field may be that due to the current from a microphone transmitter passing through a coil suitably mounted relative to the tube.

The method of controlling the output of a pliodynatron would naturally be by varying the potential of the grid. Offhand it might seem that this would either stop all oscillations (if the grid were sufficiently negative) or else let them remain at full intensity. As a matter of fact, because of the curvature of the dynatron characteristic under certain conditions, it is possible to get a control curve of the pliodynatron (grid potential-plate current) similar to that shown in Figure 175. This curve has a considerable straight line portion, and consequently between *A* and *B* thereon, it becomes possible to control the output of the tube by varying the grid potential. The actual arrangement is shown in

Figure 176. As will be seen, the circuit L_1C_1 is connected in the usual fashion for dynatrons between the plate and the battery tap point D.

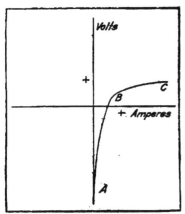

The potential variations corresponding to the speech are placed on the grid by the secondary S of the audio frequency microphone circuit transformer. The modulated output passes to the antenna circuit through the inductive or other coupling at L. In practice, radio telephony over a distance of 16 miles (26 km.) was easily accomplished with one pliodynatron; but this range could doubtless be much increased since no attempt was made at the time to get the greatest possible output or range.

FIGURE 175—Grid potential-plate current characteristic of a pliodynatron.

A system of radio telephonic control involving both an Alexanderson alternator for the direct generation of the radio frequency energy and one or more pliotrons for the modulation and control thereof is shown in Figure 177. As will be seen, the radio frequency alternator is coupled inductively to the antenna by the coils L_1 and L_2. The antenna is tuned by the variable inductance L,

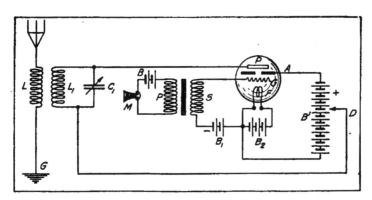

FIGURE 176—General Electric Company-Hull pliodynatron radiophone transmitter.

and the top H of the tuning inductance is the point of highest potential within the station building. (Of course, the highest potential produced by the set is at the relatively inaccessible top of the antenna.) The filament of a large pliotron is connected to the ground, and the plate of the pliotron to the point H at the top of the tuning inductance. If the filament is heated by alternating current, the mid-point of the step-down transformer secondary whereby this is accomplished is connected to ground thus equalising the thermionic current in all parts of the filament as much as possible (as indicated in the description of Figures 74 and 75, page 80). If the grid of the pliotron is kept at a very negative potential, the effect on the antenna energy will be practically nothing. As the grid becomes less negative, the pliotron permits increasingly more radio frequency current to pass through in rectified half cycles,

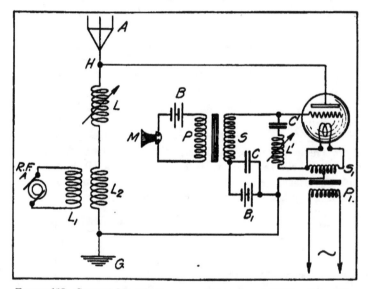

FIGURE 177—General Electric Company-Alexanderson-White alternator-pliotron radiophone transmitter.

thus withdrawing energy from the antenna. In other words, the output of the alternator either passes into the antenna system or into the pliotron bulb. It is found by experience that the fact that the pliotron absorption takes place only for half cycles does not affect this conclusion.

It will be noted that the grid is normally maintained at a negative potential by the battery B_1, which battery is shunted by the condenser C which acts as an audio frequency by-pass. The secondary of the audio frequency transformer S is also included in the grid circuit, and thus the grid potential is also caused to vary in accordance with the speech forms. In thus controlling the antenna energy by the pliotron, a curious difficulty arises. The impressed radio frequency plate potentials are quite high, and there is capacitive coupling between the plate and grid *within the bulb* since these metallic masses are, in effect, the parallel plates of a condenser. In consequence, there will be induced smaller, though still troublesome, radio frequency potential variations on the grid. During the positive half cycle, a positive potential is induced on the grid which may be much larger than the potential supplied to the grid from the telephone transmitter. This action, therefore, prevents control. This would render the system inoperative, but the effect is avoided by the introduction of the radio frequency short-circuit $L'C'$ between the grid and the filament, whereby no radio frequency potential variations can occur on the grid.

Another form of the same general type is shown in Figure 178. In this form also the control system of energy absorption by the pliotron is

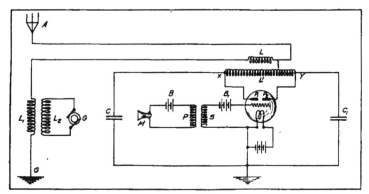

FIGURE 178—General Electric Company-Alexanderson alternator-pliotron control
radiophone transmitter.

used, but in addition an appropriate radio frequency transformer LL' is provided. This raises the applied voltage to a value most suitable for the pliotron actually available. In other words, instead of absorbing a given amount of energy at low voltage and high current it is absorbed at high voltage and low current. Furthermore, there are provided two plates P_1 and P_2 of the pliotron so that absorption occurs during both half cycles. The actual appearance of the step-up transformer which has been used experimentally is given in Figure 179. It is an open core auto-transformer consisting of a number of flat coils hung on wooden rods. One or two of the central sections are tapped to form the primary and the whole set of coils, terminating at wires X, Y constituted the secondary. Special forms of end shields designed to prevent excessive corona and break-down are mounted at the ends of these sets of coils. The exact mode of operation of this transformer is described in "Proceedings of the Institute of Radio Engineers," Volume 3, number 2, page 138.

This transformer has very low losses, so that it becomes possible to transform from 250 volts to 100.000 volts at 100,000 cycles. Under these conditions, the inductance of the transformer system was such that 2 amperes appeared at the center of the secondary winding. A study of the action of this transformer shows that if the decrement of the secondary tuned circuit be increased (by the pliotron) from its normal value of about 0.008 to about 0.8, the effective impedance of the system will

FIGURE 179—Step-up transformer for radio-frequency
high voltage transformation.

increase from 125 ohms to 12,500 ohms. One unusual characteristic of this method of varying the radio frequency resistance of the antenna, by inserting therein the primary of a transformer the secondary circuit of which contains a pliotron, is that maximum secondary current naturally corresponds to minimum antenna current.

This system of control enabled radiophone communication between Schenectady and Pittsfield, a distance of 50 miles (80 km.), a small 2 K. W. alternator running at 90,000 cycles being used as the source.

Absorption systems of these types may be used as direct, median, or inverted modulation systems. That is, we may arrange so that, when no speech is taking place and the microphone circuit resistance is therefore a maximum, the maximum current flows in the antenna; this current to be suitably diminished by modulation whenever speech begins. Or the current in the antenna may center about a median value corresponding, for example, to half-energy. Or finally, the antenna current corresponding to the undisturbed microphone may be practically zero, to increase by modulation at the beginning of speech. This inverted modulation would seem preferable on the basis of reduced radiation during inactive periods. However, only the median modulation will, in general, give satisfactory articulation.

One interesting point remains to be mentioned in connection with all modulation systems. If a 100,000 cycle sustained wave be modulated by a 1,000-cycle note, both theory and practice agree as to the propriety of regarding the modulated wave as the resultant of *three* separate waves: namely, one corresponding to the frequency of 100,500, one corresponding to the frequency of 99,500, and one corresponding to the frequency of 100,000. All three, being physically present, are detectable with a wave meter, and this has a certain bearing on the selectivity in radio telephony, particularly at very long wave lengths, corresponding to low radio frequencies.

CHAPTER VIII.

(h) FERROMAGNETIC CONTROL SYSTEMS.

We pass now to a highly valuable group of control systems wherein
the magnetic properties of the iron cores of inductances are utilised.
They depend on the following principle. The permeability of iron is not
constant; that is, the magnetic flux or induction through the iron core
of an inductance is not directly proportional to the applied magnetising

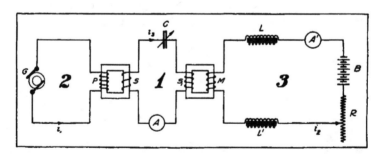

FIGURE 180—Control of radio frequency current in resonant circuit by variation
of magnetisation of iron core of inductance.

force (in ampere turns) but varies in the manner which was discussed
in the description of Figure 107, page 110, though in connection with a
different application to frequency changers. In consequence, the inductance of such a coil is dependent on the current. Starting with very
small magnetisation, the permeability rapidly increases to a maximum
and then slowly drops till it reaches the value unity for very high flux

densities. Similarly, beginning with a small current through an iron core inductance, the inductance of the coil first rises rapidly, and then drops slowly. This point will be illustrated hereafter.

We shall consider only two radiophone systems based on this principle, since these two are the only ones in actual use at present. They are the system of the Telefunken Company, as devised by Dr. Ludwig Kühn and others, and the General Electric Company's system, as devised by Mr. E. F. W. Alexanderson.

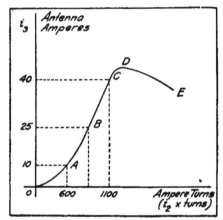

FIGURE 181—Control characteristic of resonant circuit containing iron core inductance of variable magnetisation.

Dr. Kühn was led to work out the first mentioned system by his failure in 1912 to control directly approximately 7 kilowatts of radio frequency energy by 72 microphones! The first circuit devised by him is shown in Figure 180. Here circuit 2 contains the radio frequency alternator G and the primary P of an ordinary transformer. We shall call the current in this circuit i_1. The next circuit, 1, contains the secondary S of the same transformer, an iron core inductance S_1, a tuning condenser C, and the ammeter A. In this circuit we have the current i_3. Circuit 3 contains the battery B, a variable resistance R, the ammeter A', two choke coils, L and L', to prevent radio frequency currents flowing in this circuit by induction from S_1 to M, and the magnetising coil M. If circuit 1 be tuned to resonance and then circuit 3 be closed, the inductance of S_1 will be changed because of the change in permeability of the iron core. In consequence, the current in circuit 1 will drop as the direct current in circuit 3 is increased. Conversely, we might arrange to have full resonance current in circuit 1 with a moderate direct current flowing through M, and in this case diminishing (or increasing) the direct current would cause the alternating current in circuit 1 to drop. This, then, is a system whereby the radio frequency current in circuit 1 may be caused to follow variations in the current in circuit 3.

The control characteristic of a somewhat improved system of this type shown in Figure 183, is given in Figure 181. Vertically is plotted the radio frequency current in the antenna and horizontally the magnetising force (i. e., the product of amperes and turns). It will be seen that the control is linear between point A (corresponding to 10 amperes antenna current) and C (corresponding to 40 amperes). A

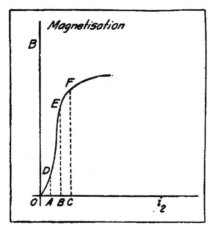

FIGURE 182—Iron magnetisation curve.

change in ampere turns of 1,100 - 600 or 500 is necessary to effect this change in antenna current. The reason why the curve of Figure 181 bends at C is shown in Figure 182, which is the magnetisation curve of the iron core of the controlling inductance. It will be seen that the control must be much more effective for magnetising currents lying between the value OA and OB than for values lying between OB and OC, since the difference between BE and AD is considerably greater than the difference between CF and BE.

A control system of the type shown in Figure 180 will be most effective under the following conditions. (By effectiveness is meant a maximum change in the alternating current i_3 for a given change in the direct current i_2.)

1. When a given amount of change of direct current energy in circuit 3 causes the greatest possible change in the inductance of S_1;

2. When the damping of circuit 1 is a minimum, so as to give sharp resonance phenomena;

3. When the couplings between the various circuits are suitably adjusted;

4. When the ratio of the continually present inductance in the circuit 1 to the variable component of inductance in that circuit is a minimum. Some of these requirements are incompatible with each other. For example, requirements 2 and 4 may easily conflict. A rational compromise must then be effected. As far as requirement 2 is concerned,

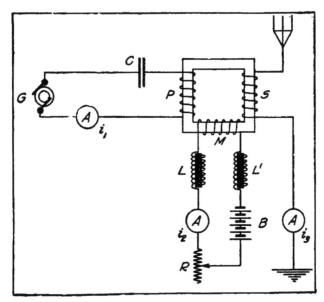

FIGURE 183—Telefunken Company-Kühn system of antenna current control.

this will require the use of very thin sheets of special iron as the core of the inductance S_1. In fact, sheets 0.001 inch (0.02 mm.) to 0.002 inch (0.04 mm.) are recommended for this use.

An improvement on the system of control shown in Figure 180 is given in Figure 183. It will be seen that in this case the magnetising coil M controls the inductance P of the radio frequency alternator circuit and also the inductance S in the antenna circuit. Consequently both circuits may be detuned and the resulting change in antenna current for a given change in magnetising current i_2 will be considerably enhanced.

A further study of Figure 181 will indicate that the portion AC of the control characteristic should be as long as possible, and as straight and steep as possible. It was feared at first that working with iron core inductances in the radio frequency circuits, hysteresis effects might distort the speech so as to make it unrecognisable. Experiment, which is an unfailing criterion in such matters, demonstrated conclusively that this fear was groundless. Care must be taken, however, not to exceed the limits of antenna current imposed by the straight line portion of the control characteristic. For an actual characteristic given in Figure 181, the change in antenna energy between 10 amperes and 40 amperes would be 5.4 k.w. with the antenna used. In use for radio telephony, a somewhat more limited range of control was used.

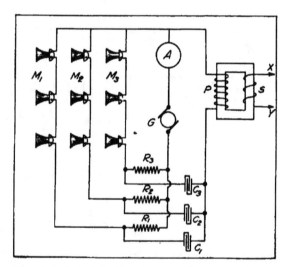

FIGURE 184—Telefunken Company-Kühn system for utilising microphones in parallel on direct current.

In order to secure the necessary control speech current, Dr Kühn devised the series-multiple arrangement of microphones shown in Figure 184. This is considered at this point instead of under "Microphone Control Systems" because no radio frequency energy is supposed to pass through the microphones and they control only indirectly through a ferromagnetic inductance. Each of the microphone banks M_1, M_2, M_3 is fed from the same generator G and through its individual large resistance R_1, R_2, R_3. Across each microphone is shunted the primary P of the telephone current step-down transformer and the corresponding one of the three condensers C_1, C_2, C_3. The output is taken from XY.

The action is simple. Whenever the microphone resistance increases, the current through its series resistance remains nearly constant but the current through it diminishes. The excess current tends to find its way through P and the corresponding condenser. This arrangement of microphones is easily seen to be stable. The telephone transformer PS must be carefully designed. In practice it is a 10-to-1 step-down transformer with a total iron path of about 13 cm. (5 inches). The primary and secondary volt-amperes are nearly equal, the leakage, resistance, iron losses, and magnetising current being all reduced to a minimum.

It will be noticed from Figure 183 that there is a marked tendency to induce radio frequency currents in the magnetising circuit including M, since M is, in effect, the secondary of a transformer of which P is

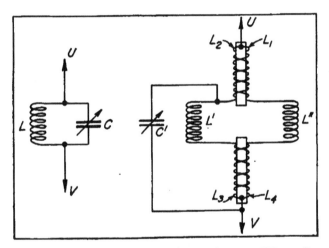

FIGURE 185—Ordinary and Telefunken Company-Kühn radio frequency choke systems.

the primary. Drastic means must be taken to avoid this because of the damage to the battery and microphones which would be done and the loss of output energy resulting. In Figure 183 ordinary iron core choke coils are indicated as the means whereby the radio frequency currents are choked off, but this means would almost always be entirely insufficient. The distributed capacity of such a coil would cause it to interpose but little impedance to the radio frequency current, in general. A more usual means is by the use of the loop circuit shown in the left of Figure 185. As is well known, the reactance of such a loop measured between the points U and V becomes infinite at the frequency for which the loop is resonant, provided there are no losses in L and C. Even if there are small losses in L and C, the impedance will become very high. An improved method whereby unusually high impedances can be secured by the coils used in practice is shown in the right hand portion of Figure 185. Here L_1 and L_2 are two coils wound in opposite directions on the same core (not of iron). L' and L'' are small inductances widely separated from each other. L_3 and L_4 constitute a double coil similar to L_1 and L_2. The tuning condenser C' is inserted as shown. For the audio frequency telephone currents, L_1 and L_2 form a system of very

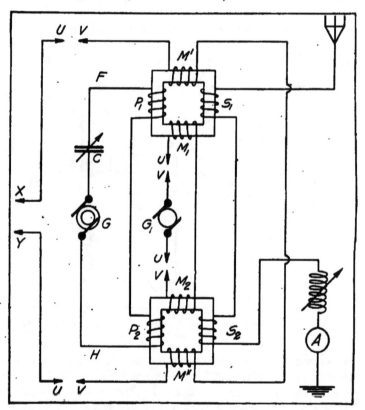

FIGURE 186—Telefunken Company-Kühn radiophone transmitter, 1913.

low inductance as do L_3 and L_4. In fact, the inductance between U and V for telephone currents is only 20 microhenrys in one practical instance. On the other hand, for the radio frequency currents the impedance is extremely high.

The latest and most improved pattern of these radiophone sets is

FIGURE 187—Telefunken Company 10 k.w. alternator-frequency doubler radiophone transmitter.

shown in outline in Figure 186. As will be seen, the generator G of radio frequency energy is placed in a tuned circuit including C and the primaries P_1 and P_2 of the frequency changers. (A description of these frequency changers has already been given in connection with Figures 107 through 111, page 110.) The secondaries S_1 and S_2 of the frequency changers are in the antenna circuit in series with a necessary tuning inductance. The direct current generator G_1 is arranged to supply the direct current magnetisation of the frequency changers by the coils M_1 and M_2. The two gaps in the circuit of this generator at UV are supposed to be filled with choke systems such as those of Figure 185, the lettering corresponding. The telephone control current produces *changes* in the otherwise constant magnetisation of the frequency changer cores in passing through the coils M' and M''. The telephone currents originate in the gap XY which corresponds to the terminals XY of Figure 184, the remainder of the microphone system being omitted from Figure 186 for the sake of simplicity. For the same reason, the choke systems at points UV in the telephone control circuit are only indicated. It will be noted that the system here shown differs from the simpler system of Figure 183, not only in the use of the frequency changers but also in the separate constant direct current magnetisation and separate telephone control current magnetisation. Instead of having only one set of frequency changers, the terminals FH may themselves be the output terminals of one or more frequency changers these being placed where the generator G is indicated.

An actual radiophone set of this type is shown in Figure 187. This set is supposed to be run from 110-volt direct current mains. A motor drives the 10 K.W., 10,000 cycle alternator, which is similar to that shown and explained in connection with Figures 112 and 113, page 114. The frequency may be raised in four steps to 160,000 cycles corresponding to 1,880 meters wave-length. In the middle of the top crown panel is a control stroboscope for watching the telephone control. This device is a small neon or carbon dioxid vacuum tube rapidly rotated by a small motor. It is connected through a small capacity to a high potential point of the antenna system, and when there is sustained radiation, a uniform circular band of light caused by the rotating tube indicates this fact. If a musical sound affects the microphone transmitter, the circular band of light is broken into narrow radial bands, and the relative brightness of the center of the bands and the darkness of the middle of the space between them indicates very roughly the completeness of the modulation. However, such instruments are far from quantitative, being at best rough indicators. The top row of instruments are respectively for the direct current supply voltage, the current supplied a special small motor, the excitation (field) direct current, the 80,000 cycle telegraph control key circuit, and a 0-to-40 ampere antenna ammeter. The second row of instruments are the large motor ammeter, the "magnetising" current, an 80-ampere ammeter for the 10,000 cycle output circuit, a 10-ampere ammeter for indicating the alternating current from the microphone transformer (corresponding to XY of Figure 184), and the antenna current ammeter for telephony. The lower left panel carries the large driving motor switch, the magnetising current switch, and control switches and fuses for the stroboscope and

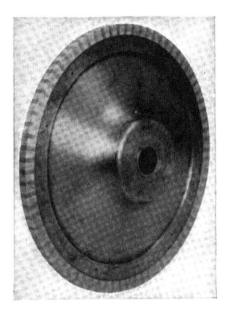

FIGURE 188—General Electric Company-
Alexanderson telephone control
relay; rotor.

the ventilating fan motor. The center lower panel carries the key relay (for telegraphy), a field rheostat, the frequency regulating device of the musical tone producer for ordinary telegraphy (simulating a spark station), and the wheels which control the tuning inductances (variometers) of the 10,000 cycle circuit, the 40,000 cycle circuit, and the 80,000 cycle key circuit. On the right hand lower panel are the control wheels of the inductances in the 20,000, 80,000, and 160,000 cycle (antenna) circuits, the frequency meter for musical tone telegraphy, and 8 or 10 microphones suitably arranged. The desk carries the telegraph key and the bottom panel to its left the motor starters and regulators.

The entire outfit can put about 6 kilowatts into the antenna at 1,900 meters for telegraphy and several kilowatts for telephony. Some figures given by Dr. Kühn indicate that a microphone output of about 4 watts (or 20 volt-amperes), corresponding to a control alternating current of 8 amperes through the 30 turns of the 40 microhenry control windings on the final transformers, suffices to control several kilowatts, the energy amplification being as great as 1,000.

With a set similar to that shown, using the Nauen antenna and at a wave-length near 5,000 meters, speech was transmitted from Berlin to Vienna, a distance of 340 miles (550 km.), the received words hav'ng

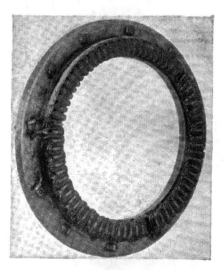

FIGURE 189—General Electric Company-
Alexanderson telephone control relay;
stator (field and armature).

an audibility of 100. Professor Kann, listening at Vienna, stated that there were unusually heavy atmospheric disturbances. The speech was clear but the vowels were emphasized while the consonants seemed sometimes to be almost missing. On the other hand, singing was faultlessly transmitted. How far these effects were due to the heavy strays and how far to iron distortion of the speech forms is not stated.

We consider next a further development of the ferromagnetic control systems, namely Mr. Alexanderson's magnetic amplifier as designed for the General Electric Company. Prior to considering this device, the parent idea from which it sprang will be given. This was a so-called "telephone relay." It was a moderately high frequency

alternator of the inductor type the field of which was varied by the
speech current. In consequence the output of the machine was similarly
modulated. The rotor of the machine is shown in Figure 188. The
iron teeth had to be laminated because of the variations in the field produced by the speech current. This was a serious limitation of the
machine. The stator of the machine is similarly shown in Figure 189.
The zig-zag winding of the alternator around the teeth is clearly visible.
Underneath this winding are field windings. To avoid the limitation
mentioned above, the modern magnetic amplifier was invented, this
being a device which has practically the same effect, when placed across
the terminals of a radio frequency alternator, as would speech variation
of the field thereof.

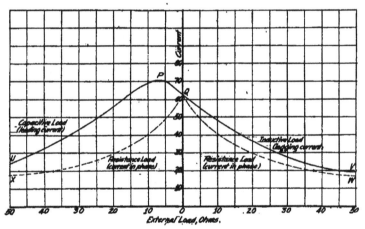

FIGURE 190—Load characteristics of 50 K. W., 50,000 cycle Alexanderson
alternator.

Let us consider first the operating characteristics of an Alexanderson
radio frequency alternator, namely the 50 k.w., 50,000 cycle machine
shown in Figure 124, page 124. These characteristics are shown in
Figure 190. It will be seen that if 50 ohms of external resistance are
shunted across the machine terminals, the current (at the point W)
will be 17 amperes. As this load resistance is diminished, the current
rises along the dashed curve to the point Q. This corresponds to zero
resistance across the alternator and to a current flow of 63 amperes.
This current, since the load is a pure resistance, is in phase with the
voltage (neglecting machine impedance), and the curve has been repeated
symmetrically to the left of the current axis by the curve QX. If various
reactances (in the form of resistance-free inductance) are placed across

4

the terminals of the machine, the curve QV is obtained for the relation between current and external reactance in ohms. Thus the point V corresponds to 50 ohms reactance or 0.00016 henry at 50,000 cycles. The current in this case lags behind the electromotive force since the load is inductive. If 50 ohms of capacity reactance, which corresponds to 0.064 microfarad at 50,000 cycles, and is indicated at the point U, be placed across the machine, a current of 24 amperes is obtained. As the external capacitive reactance is diminished, the current increases, reaching a maximum of 71 amperes at 8 ohms corresponding to 0.40 microfarad at 50,000 cycles. On leaving the point P of maximum current, with diminishing external reactance and corresponding external

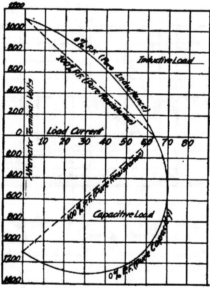

FIGURE 191—Load characteristics of 50 K. W., 50,000 cycle Alexanderson alternator.

capacity load, the curve drops again to the point Q. In the portion QPU of the curve, the current leads the voltage since the load is capacitive. It will be seen that the curve $UPQV$ is nothing more than the resonance curve of the system made up of the alternator armature and the external load. Since the capacity reactance for resonance is 8 ohms, the inductive reactance of the alternator armature must have the same numerical value. Consequently the inductance of the armature must be approximately 26 microhenrys at 50,000 cycles, an interestingly low value.

The same material is plotted in another fashion (based on the curves given by Mr. Alexanderson in an earlier publication) in Figure 191. The curves given differ from those of Figure 190 only in that alternator terminal volts and load current are plotted instead of external impedance and load current. It may be noted that the 0 per cent. power factor curve is that of a *pure* inductive or capacitive load; that is, one which is resistance-free. In the same way, the 100 per cent. power factor resistance curves are with a load consisting of *nothing but resistance*. While not quite so clearly visible, the resonance phenomenon is indicated here also.

The general arrangement of the magnetic amplifier in its simplest form are represented in Figure 192. The nature of the iron structure is sufficiently indicated. Coils L_1 and L_2 are wound over the two middle

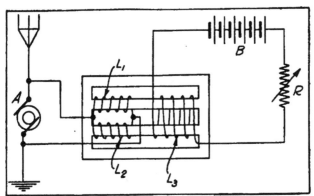

Figure 192—General Electric Company-Alexanderson magnetic amplifier (shunt connected to alternator, multiple connection of coils).

cores, connected in parallel, and the combination shunted across the radio frequency alternator A. (Coils L_1 and L_2 are placed in parallel rather than in series since theory and experiment agree in predicting a more effective control by such connection.) It will thus be seen that the iron core inductance L_1 L_2 is placed across the alternator terminals. If this inductance is varied by any means, the right hand curve of Figure 190 will indicate the current variation through the inductance. Consequently the antenna current will also vary in the opposite sense, and a marked degree of antenna current control would be thus obtained. The mode of varying the inductance of coils L_1 and L_2 is also shown clearly in Figure 192. It is by means of the coil L_3 through which passes a direct current from the battery B which current can be suitably varied by the control resistance R. It will be seen that L_3 is wound over the cores of both L_1 and L_2 and thus there will be therein no radio frequency induction from the latter. This is important, and constitutes a marked advantage of Mr. Alexanderson's magnetic amplifier over the device used by the Telefunken Company and shown in Figure 186. The actual appearance of the amplifier is given by Figure 193. The magnetising control coils L_3 are indicated as in Figure 192. The two sets of coils corresponding to L_1 and L_2 are also indicated. The coils L_1 are partly hidden by the cross piece. It may be mentioned that a number of further designs of more advanced character have been adopted recently

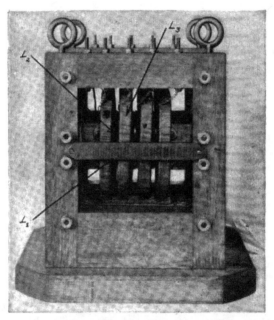

FIGURE 193—General Electric Company-Alexanderson "magnetic amplifier" radiophone control.

for the magnetic circuits of the amplifier, but the principle remains unchanged.

The actual behavior of the amplifier is well represented by Figure 194. This shows the impedance of the amplifier, expressed in ohms, plotted against the radio frequency current passing through it for various direct currents through the magnetising coils L_3. It will be seen that for no magnetisation (curve ABC) the impedance varies from 32 to 70 ohms between 60 amperes of radio frequency current and 20 amperes. With 0.7 ampere d. c. magnetisation, the variation is somewhat in the opposite sense, namely between 27 ohms and 15 ohms for

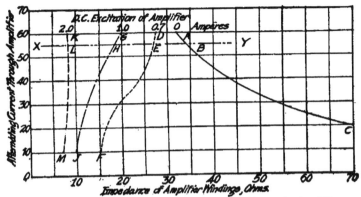

FIGURE 194—Characteristics of Alexanderson Magnetic Amplifier.

the variation of radio frequency current between 60 and 10 amperes. For 2.0 amperes magnetisation current, the impedance of the amplifier remains nearly constant around 8 ohms for the same extreme variation of radio frequency current through it. Considering the line $ADGK$, it is clear that with 60 amperes radio frequency passing through it, the amplifier impedance changes from 32 ohms to 8 ohms as the direct current magnetisation is increased from 0 to 2.0 amperes. Similarly, at 55 amperes radio frequency current through the amplifier, corresponding to line XY, a somewhat wider variation is obtained. It is thus perfectly

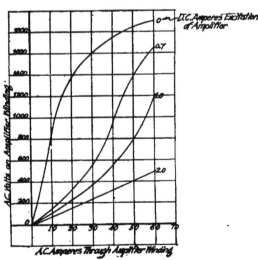

FIGURE 195—Characteristics of Alexanderson magnetic amplifier.

clear that the amplifier is a markedly effective device as a variable impedance for radio frequency currents. The same data as that represented in Figure 194 is given in different form in Figure 195, which gives the corresponding voltage-current curves of the amplifier. It will be seen that the current through the amplifier at 400 volts impressed radio frequency may be varied from 5 amperes (for no d.c. magnetisation) to 50 amperes with 2.0 amperes magnetisation. At 1,200 volts applied radio frequency, a current variation of 15 to 60 amperes (i. e., from 18 to 72 kilovolt amperes) is obtained with a variation of the d.c. magnetisation of only 1 ampere, a full illustration of the usefulness of the device.

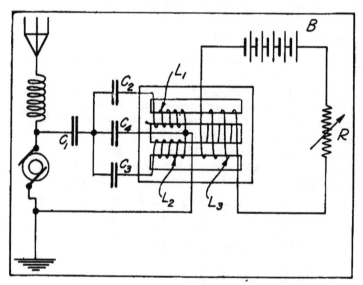

FIGURE 196—General Electric Company-Alexanderson magnetic amplifier, with series, short-circuiting, and shunt condensers.

In order to secure the maximum linear control and to prevent certain undesired effects, several condensers are inserted into the amplifier circuits as shown in Figure 196. The first of these is the series condenser C_1 which is placed between the high potential point of the alternator and a point leading to one side of the amplifier (through several other condensers to be considered below). The effect of the series condenser on the stability of operation of the amplifier and otherwise is illustrated in Figure 197. This curve shows the current flowing from the alternator as ordinate plotted against the external impedance in ohms. The dot-and-dash curve shows the effect of a purely inductive load and is the same as curve QV of Figure 190. The dashed curve for the amplifier alone is not far from curve ABC of Figure 194. To the left of the vertical axis is drawn the corresponding curve of the constant impedance of 8 ohms, this being a capacity of 0.33 microfarad. The curve in question is the vertical line marked 0.33 μf. Inserting such a condenser at C_1 will give the resulting curve to the right marked "amplifier + 0.33 μf." This curve represents a stable state of affairs. At the extreme left, the vertical dashed line shows the constant impedance of 48 ohms corresponding to a series condenser of 0.067 microfarad. The curve marked "amplifier + 0.067. μf." is the result of using this series condenser and has an unstable portion to the left of the vertical axis. This corresponds to an *increase* of current with an *increase* of impedance across

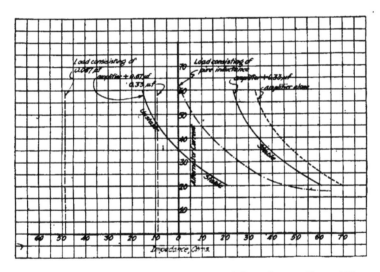

.FIGURE 197—Effect of series condenser on stability of magnetic amplifier.

the alternator. The same effect is shown in Figure 198 in different form, this figure representing alternator terminal voltage vertically and current through the amplifier and condenser horizontally. The curve marked "amplifier + 0.33 μf." is a rising curve practically throughout, whereas the curve for the "amplifier + 0.125 μf." shows a falling portion corresponding to *increasing* current with *diminishing* voltage. This is what we have called a condition of "negative resistance" such as is experienced, for example, in the Poulsen arc. Accordingly, this unstable region is unusable and may lead to self-excited oscillation in the amplifier system, which is a normally undesirable condition.

The effect of the series condenser C_1 in Figure 196 is to give a great increase in the sensitiveness of the system and also to give a linear control characteristic. The control characteristics for several values of the series condenser are shown in Figure 199, which should be carefully compared with Figures 126 through 128, page 128. Curve A of Figure 199 shows the relation between *antenna current* in the arrangement of Figure 196 and the d.c. *amperes in the magnetising coil* of the amplifier (L_8 in Figure 196). It is the real control characteristic of the system

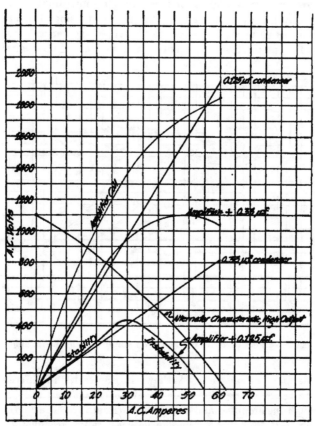

FIGURE 198—Effect of series condenser on stability of magnetic amplifier.

when used for radio telegraphy and telephony. Curve B, obtained with a series condenser of 0.33 microfarad shows practically complete and linear modulation except for excessive control to the left of the point B, this corresponding to greater amplifier magnetising currents than about 2.8 amperes. To the left of this point the control reverses as indicated in Figure 5 which corresponds to this case. Between Y and Z of that figure we are working on portion BD of curve B of Figure 199, but between X and Y of that figure we are working on the reversed portion of curve B of Figure 199. It need hardly be said that in practice this condition can be and is easily avoided. A smaller series condenser of 0.125 microfarad gives control characteristic C of Figure 199. This is a steeper control than those of the preceding cases, but it is incomplete and therefore not chosen. Some study will convince the reader that these control curves are closely related to

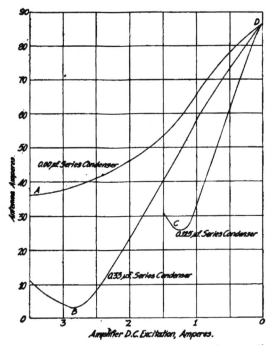

FIGURE 199—Control characteristics of magnetic amplifier.

inverted resonance curves in a system having moderate effective resistance and iron losses.

The second condenser considered, namely, C_4, in Figure 196, is known as the shunt condenser. Its function, according to Mr. Alexanderson, is to cause the amplifier to take a leading, instead of a lagging current at low excitations and to increase the sensitiveness of the arrangement. According to Mr. Louis Cohen, it may rather be treated as forming with the amplifier a loop circuit the impedance and effective resistance of which change very markedly near a resonant frequency.

The third condenser (actually the two condensers C_2 and C_3) is known as the short-circuiting condenser. It will be noticed that there is a closed circuit $L_1C_2C_3L_2$ in which audio frequency currents may be induced if telephonic currents flow in the control winding L_2. These would be short-circuited except for the two condensers just mentioned, and the control would become ineffective. The condensers C_2 and C_3 are so chosen that their audio frequency reactance is very high while their radio frequency reactance is quite low. In this way the radio fre-

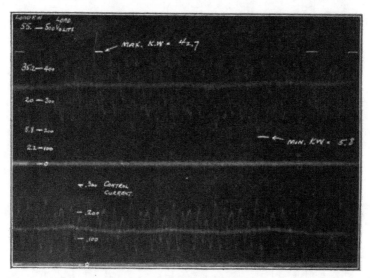

FIGURE 200—Oscillograms showing controlling telephone current and controlled antenna output of General Electric Company-
Alexanderson 50 k.w. alternator and
magnetic amplifier.

quency currents may still flow practically undeterred through the amplifier coils while the audio frequency currents are almost entirely prevented from doing so.

The combination of these various condensers gives a high degree of control. Experiment shows that the amplification, defined as the radio of (maximum antenna kilowatts minus minimum antenna kilowatts) divided by (effective kilovolt-amperes in the control circuit), varies from 100-to-1 to as much as 350-to-1. This is under the linear conditions necessary for control in telephony. The perfection of the control is well illustrated in Figure 200. The lower oscillogram shows the control current in L_3 in Figure 196, while the upper curve shows the antenna kilowatts. It will be seen that a variation of control current of 0.2 ampere changes the antenna kilowatts from 5.8 to 42.7, a variation of nearly 37 kilowatts! The Author believes this to be the largest amount of radio frequency (or other) energy ever controlled by a telephone transmitter. It will be noted that the antenna kilowatt curve is inverted relative to the control current curve, the peaks in one corresponding to the crests in the other. This is the result of the control characteristic of Figure 199, which shows that *large* antenna currents correspond to *small* control currents and vice versa.

(I) COMPARISON OF CONTROL SYSTEMS.

The choice of modulation control system will depend markedly on the output of the radiophone transmitter and, to a less degree, on the type of installation, i. e., ship or shore station, fixed or portable outfit.

For low-power sets, the placing of the microphone directly into the antenna (as illustrated in Figure 10, page 22) is a simple solution and one that is economical of apparatus. It is not, however, economical of energy since the microphone resistance for most efficient operation must be equal to that of the remainder of the radiating system. This condition necessarily involves the loss of half of the available radio frequency energy in the microphone. To some extent this loss may be avoided by the use of one of the circuits shown in Figure 130, page 133; whereby the microphone is more fully utilised in that the changes in its resistance vary a number of the electrical constants of the associated circuits and thus produce greater proportionate changes in the antenna current.

For moderate power sets, the difficulties in getting a suitable control system become quite serious. Large numbers of microphones in parallel are bulky and expensive, and tend to cause difficulty in adjustment. Heavy current microphones seldom give the highest quality of articulation and liquid microphones are not easy to build or to shield from disturbance. On shipboard their use is even less desirable than on land. It becomes necessary to use some type of control based on one-way amplifiers (such as the methods of the General Electric Company involving absorption in pliotrons as shown in Figures 177 and 178, page 178, and a number of allied methods) or else to use ferromagnetic amplifiers. These last should be so constructed that they are one-way devices so far as possible, in order that there shall be no induction of radio frequency currents in the control or microphone circuit.

For high power sets, direct microphone control is, of course, out of the question. Even the use of the normal vacuum tube amplifier in any of its modifications or modes of use seems of doubtful utility unless some very heavy output bulbs should be constructed in the future.

The most feasible methods at present seem to be those involving the control of the outgoing energy by rugged ferromagnetic amplifiers. The control energy for these amplifiers may itself be obtained by the use of vacuum tubes or of smaller ferromagnetic amplifiers. In other words a composite system depending on the use of a rugged final amplifier is desired, its control energy being derived from a more delicate amplifier which can be actuated by the small amount of microphone energy actually available.

It may be here mentioned that the difficulties of the situation are considerably increased when it is desired to control the radiophone transmitter from an ordinary wire telephone line. The power available from an ordinary telephone line is of the order of microwatts, whereas the power derived directly from the transmitter may be hundreds or thousands of times as much. The difference must be made up in the former case by at least one audio frequency amplification.

CHAPTER IX.

8. ANTENNAS AND GROUND CONNECTIONS—(a) RADIATING SYSTEMS; ANTENNA RADIATION EFFICIENCY. 9. RECEPTION PHENOMENA. (a) DETECTOR AND AMPLIFIER TYPES. (b) BEAT RECEPTION; CONSTANCY OF RADIATED FREQUENCY. (c) SELECTIVITY IN RECEPTION. (d) INTERFERENCE WITH RADIOPHONE RECEPTION. (e) TELEPHONE RECEIVERS. (f) RECEIVING APPARATUS; AUDION RECEIVERS; ARMSTRONG REGENERATIVE RECEIVERS; ULTRAUDION CIRCUIT; TELEFUNKEN COMPANY; MEISSNER RECEIVING ARRANGEMENTS; MARCONI COMPANY RECEIVER; WESTERN ELECTRIC COMPANY TUBES FOR RECEPTION; TUBE DESIGNS OF VAN DER BIJL, NICOLSON, AND HULL; GENERAL ELECTRIC COMPANY TUBES FOR RECEPTION AND AMPLIFICATION (PLIOTRONS). (g) STRAYS; BALANCED VALVE RECEIVERS; DIECKMANN CAGE; CLASSIFICATION OF STRAYS; COMPENSATION METHOD OF STRAY REDUCTION; METHODS OF DE GROOT. (h) RANGE IN RADIO TELEPHONY; OCCASIONAL RANGE; RELIABLE RANGE; ANNUAL INCREASE IN RANGE.

8. ANTENNAS AND GROUND CONNECTIONS.

(a) RADIATING SYSTEMS.

For transmission in radio telephony, much the same requirements must be met as in the case of a normal, sustained wave, telegraph station. That is, a high capacity antenna of low ohmic resistance is desired. The radiation resistance should be the chief portion of the total antenna resistance else the efficiency of radiation may be very low. This can be well illustrated by the following numerical example:

Suppose an antenna to have an effective height of 40 meters (130 feet), and that 1 kilowatt is available for the antenna circuit. Suppose further that the antenna is used successively at wave-length of 1,600, 3,200, and 6,400 meters. In the table, page 206, are given values of the probable ohmic resistance of the antenna, its ground resistance, and its radiation resistance at each of the wave-lengths. These are calculated on the basis that the ohmic resistance increases as the frequency increases

(i. e., as the wave-length diminishes), that the ground resistance diminishes almost inversely proportionately to the wave-length, and that the radiation resistance is inversely proportional to the *square* of the wave-length. The antenna current is then calculated from the total resistance, and the radiated energy and radiation efficiency in per cent.

	WAVE LENGTH		
	1,600 m.	3,200 m.	6,400 m.
Ohmic Resistance of Antenna.....	0.3	0.2	0.1
Ground Resistance	1.0	2.0	4.0
Radiation Resistance	1.0	0.25	0.06
Loading Coil Resistance.........	0.3	1.2	4.8
Total Resistance	2.6	3.65	8.96
Antenna Current	19.6	16.6	10.6
Radiated Power (watts).........	385.	69.	7.
Radiation Efficiency (in per cent.)	39.	6.9	0.7

It is clear enough, everything else remaining the same, that the shortest wave-length (nearest to the antenna fundamental) would be by far the most suitable so far as radiation efficiency is concerned. However, the absorption of the electromagnetic waves in passing over the intervening country may partially or entirely nullify this difference, and thus it may occur that by day and for overland transmission the best wave would not be the 1,600 meter wave but possibly the 3,200 meter wave. For the case mentioned, with the relatively small antenna power available, the transmission could hardly be over a great enough distance to make the longest wave given the most desirable. In other words, in every case of day transmission, there will be some wave-length for which best results are obtained because a diminution of wave-length below this most favorable value, while it would increase the radiation efficiency, would more than correspondingly diminish the freedom from absorption in the intervening space.

In constructing antennas for radio telephony, all the usual precautions as to antenna insulation, ground resistance, freedom from neighboring energy-absorbing conductors and guy wires are observed. In addition, it should be remembered that it will sometimes be necessary to use a suitable coupling between the antenna and the radio frequency generator in order that the resistance which is thus, in effect, introduced into the generator circuit shall have the most favorable value for full generator output.

A "fly-wheel effect" similar to the "inertia effect" mentioned under "Causes of Distortion in Radio Telephony," page 13, may occur in the antenna circuit. If the persistency of the antenna system is very great, i. e., if the damping is very small, the wave trains in the antenna will tend to persist at full intensity and the difficulty in getting complete modulation may become excessive. As regards this feature, which is most prominent at long waves, there is a conflict between good antenna design in the usually accepted sense as indicated above, and the design indicated to avoid the fly-wheel effect. In general, however, the compromise will be satisfactory if the fly-wheel effect is practically disregarded, except for long waves and very persistent antennas.

(b) RECEIVING SYSTEMS.

The same general considerations which were found to hold for transmitting antennas also hold for receiving antennas except that smaller antennas will in general be used. This is because of the diminished expense, because of the large static charges which readily accumulate on large antennas, and because it is easy enough to amplify the signals received on a small antenna to satisfactory values. Of course, the fly-wheel effect mentioned previously may again occur in the tuning of sharply resonant radio frequency circuits, though the Author has not experienced much trouble on that score using waves of moderate length. The present tendency seems toward the use of small antennas with sensitive receiving sets and high amplification of some sort. It seems that the ratio of signal strength to stray intensity remains reasonably constant as the size of the antenna is diminished, at least when crystal detectors or detectors of the audion type are used. For the oscillating audion, Mr. Armstrong has pointed out that this is not necessarily the case, since the oscillating audion favors weak signals compared to heavy strays to a greater extent than does the plain audion. So that, unless the beat method is used for reception, radio telephonic reception may just as well be carried on on small antennas as on large.

9. RECEPTION PHENOMENA.

(a) DETECTOR AND AMPLIFIER TYPES.

Almost all detectors have been used for radio telephony, and indeed all but the coherer type can be used. At the present time such detectors as the crystal rectifier, the audion, and the dynatron have proven to be practically usable and satisfactory. The detectors and amplifiers used in radio telephony should have a linear characteristic like that mentioned in connection with Figure 6, page 15. Otherwise there will be speech distortion of the types described in the discussion to which reference has been made. Both detectors and amplifiers should be of such sort that they are easily adjusted to maximum sensitiveness, retain this sensitiveness indefinitely, do not require frequent renewal, and are inexpensive. These requirements have not yet been entirely met.

(b) BEAT RECEPTION.

Beat reception is possible in radio telephony, and there may be used for this purpose either the normal detector with an external oscillator circuit coupled to the receiving system to produce the beats or the so-called "self-excited heterodyne" where the same vacuum tube is used at once as an oscillator, detector, and amplifier. Generally speaking, this latter arrangement, while convenient in manipulation and economical of equipment, does not utilise to the full the various properties of the bulb and is less stable and certain of adjustment than the former.

It need hardly be said that for beat reception in radio telephony extreme constancy of frequency at the transmitting end is essential. This will be evident when it is considered that radiophone reception under these conditions requires either zero beats per second (that is, equality of frequency of the transmitter and of the local oscillator at the receiver) or a beat frequency above audibility (that is, a greater difference between the transmitter frequency and the local oscillator frequency than say 10,000 cycles per second). As a matter of fact, only the first of these expedients is practically usable since the detuning of the antenna and its associated circuits in the receiver for the second case would make the reception very inefficient except on extremely short waves where a difference of frequency of 10,000 or more cycles per second is only a small percentage of the main transmitter frequency. However, it must be admitted that zero beat frequency is usually not very easy to obtain or hold as a receiver adjustment and even slight variations in transmitter or receiver oscillation frequency will then cause a drummy quality to appear in the speech and seriously impair its intelligibility.

With radiophone transmitters employing alternators, or alternators and frequency changers, very perfect speed regulation will therefore be required if beat reception is to be used. For example, working at 6,000 meters wave-length (50,000 cycles per second), a much greater speed variation than one part in 10,000 would be objectionable; and if frequency multipliers were employed in conjunction with the alternator to get the 50,000 cycles per second, even greater accuracy would be necessary. When bulb radiophone transmitters are used, the filament currents and reactions on the oscillator must be kept quite constant else there will be changes in the emitted frequency even in this case, and beat reception will not be feasible.

(c) SELECTIVITY IN RECEPTION.

There is a fairly sharp conflict between the requirement of loud signals and extreme selectivity. The first of these generally requires sensitive detectors and powerful amplifiers used with close coupling to the antenna system, while the second tends in the opposite directions. Nor does beat reception solve this problem as will be evident below. All that can be said is that a rational compromise must be effected in every case, this to be determined by the operating conditions in the neighborhood of the receiving station. Thus the amount of interference in the vicinity of the receiver is an extremely important factor in determining the amount of power required at the transmitter to cover the desired distance. This is a factor which is often overlooked in the design of stations.

There is also, particularly at long waves, a conflict between the extreme antenna persistence necessary for adequate selectivity in reception and the undesired fly-wheel effect which has been previously mentioned. This, again, must be met by compromise.

The effect of modulation on selectivity has been considered on page 181.

(d) INTERFERENCE WITH RADIOPHONE RECEPTION.

Interference from spark stations disturbs radiophone reception less than might be expected, partly because the dots and dashes constitute a more or less intermittent disturbance through which portions of the words can be heard and partly because of the resulting "assistance of context" effect. Sustained wave station interference is, however, very serious since this causes a continuous musical note by the beats with the incoming radiophone frequency and this continuous musical note cannot be tuned out either by ordinary or beat reception being a physically present phenomenon caused by two frequencies *external* to the receiving station. In the neighborhood of a large arc radio telegraphic station, this may become a very grave matter particularly if compensation waves are used by the arc station in transmission. In this latter case, there will generally be produced a long series of overtones of both the sending and the compensation waves, and there is very likely to be continuous beat interference. The Author is very much of the opinion that radiation at non-useful frequencies should not be permitted since the growth of the radio art will be much hampered thereby. Furthermore, provision should be made in all sustained wave stations to avoid the production of these series of overtones (which, it may be mentioned, are frequently not harmonics but fall at non-integral multiples of the main and useful frequency). Furthermore, bulb beat receivers should be so designed that the locally generated oscillations are without overtones.

(e) TELEPHONE RECEIVERS.

It might be expected that there would be no great difference between the various telephone receivers used in radio sets, so far as speech reception were concerned, but this is far from being the case. In addition to marked differences in intrinsic sensitiveness, the receivers show differences as to the extent to which they distort speech and the relative extent to which they respond to the sudden shocks caused by heavy strays. Generally speaking, the receivers with diaphragms of moderate thickness give good articulation, moderate sensitiveness, no inordinate response or "singing" when stray impulses are received, and are robust. More sensitive receivers with very light diaphragms tend to give "tinny" speech and more than proportionate response to impulses.

A number of other types of receivers besides the usual electromagnetic type have been suggested. Thus Messrs. Fessenden, and, later, Ort and Rieger have built electrostatic receivers. These are nothing more than a condenser one or both sets of plates of which are movable. The electrostatic forces developed as the difference of potential between the plates changes will cause minute movements of the plates and consequent sound. Sometimes an auxiliary potential is kept constantly on the plates and they are under considerable tension, this being found to increase the sensitiveness greatly. Such an arrangement, though it approaches the usual receiver in sensitiveness is not particularly convenient and has not found favor in the commercial radio field.

Mr. Fessenden has further developed and used a receiver based on alternating current repulsion between two coils of wire each carrying the same current, or a current of nearly the same frequency. The construction of the device was simple. Two flat spirals of thin wire were placed parallel and near to each other, and the incoming current passed through both, or else through one of them with a locally generated radio frequency current passing through the other. While the device was operative, it did not find favor in the radio field, and is not used in practice at present.

(f) RECEIVING APPARATUS.

The first receiver we shall consider is that shown in Figure 201. It is the usual audion used as a detector. Incoming radio frequency energy causes radio frequency potential differences at the terminals of the secondary tuning condenser C_1. Consequently alternating current tends to flow in the grid-to-filament circuit, $C_1 C_2 GF$. However, since the grid-to-filament has unidirectional conductivity only, the grid gradually accumulates a larger and larger negative charge, which charge cannot escape through L_1 to the filament because of the grid condenser C_2. In consequence of the increasingly negative potential of the grid, the current in

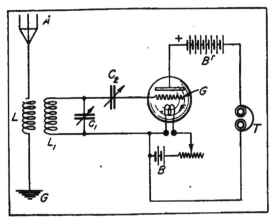

FIGURE 201—Normal audion receiver.

the plate circuit diminishes. If the signals cease, the grid leakage
(through the condenser C_2, through the glass supports of the grid, and
because of any residual positive ionisation due to gas molecules in the
space between the grid and filament) will speedily bring the grid potential back to normal and the plate current will then increase to its usual
value. As a result of this action, variations in the incoming radio frequency currents, such as occur in radio telephony, will be approximately
followed by changes in the plate current of the audion. A supplementary
resistance may be shunted across the grid condenser so as to increase grid

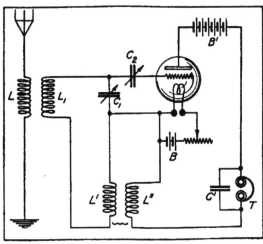

FIGURE 202—Armstrong regenerative circuit for radio
frequency amplification.

leakage and improve the fidelity of reproduction of speech in the plate circuit. This will generally diminish the audion sensitiveness. The resistance used in practice for this purpose are pencil lines, graphite rods, or liquids (e. g., xylol), and have values ranging from a few thousand ohms to several megohms. In addition, high vacuum three-electrode tubes can be used as detectors by virtue of the curvature of the plate current-grid voltage characteristic.

As explained in connection with Figure 79, page 85, Mr. E. H. Armstrong has devised a number of methods of using the audion as a regenerative relay by coupling the plate and grid circuits. Such an arrangement adapted for telephonic reception and giving radio frequency amplification is represented in Figure 202. As will be seen, the grid

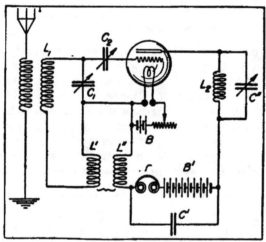

FIGURE 203—Armstrong regenerative circuit for radio
frequency amplification with plate circuit tuning.

circuit $L'C_1L_1$ is coupled to the plate circuit by means of the inductive coupling $L'L''$. Armstrong has found in bulbs used by him (high vacuum bulbs) that the regenerative amplification obtained was fifty-fold in energy or about 7 times in audibility (as audibility is usually defined, namely, as directly proportional to the current through the telephone receivers). It will be noted that the telephone T is shunted by the condenser C', the purpose of which is to permit the passage of the radio frequency current while forcing the audio frequency currents of the signal to pass through the receivers.

An improved arrangement, also due to Armstrong, is shown in Figure 203. Here, in addition to the regenerative coupling between the plate and grid circuits, we have tuning of the plate circuit by means of

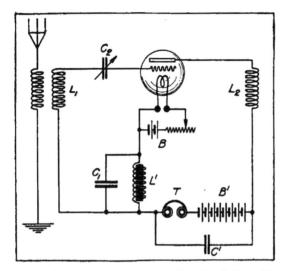

FIGURE 204—Armstrong regenerative circuit for radio
frequency amplification.

the inductance L_2 and the condenser C''. As before, the receivers T and
the plate battery B' are shunted by the by-pass condenser C'. Another
interesting modification is given in Figure 204. Here the coupling is
secured by means of the large inductance L' and the capacity C_1. The
details of this circuit together with the detailed explanations of the
various circuits here outlined and similar circuits can be obtained in the

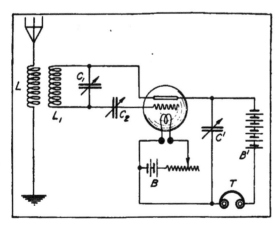

FIGURE 205—de Forest ultraudion receiver.

"Proceedings of The Institute of Radio Engineers," for September, 1915. It need only be mentioned here that it is recommended that the inductances in the plate and grid circuits be large and the capacities small.

For the sake of completeness, we include here as Figure 205 the de Forest ultraudion circuit which has previously been explained in connection with Figure 86, page 91. The modified ultraudion circuit having grid and plate circuit coupling by means of the so-called "tickler coils" is shown and explained in connection with Figure 87.

The actual appearance of a de Forest assembled audion and ultraudion receiving set is indicated in Figure 206. The tubular audion

FIGURE 206—de Forest Company receiver, 1914.

is mounted at the left with its carbon sector potentiometer (for obtaining a continuously variable plate potential) to the right of the supporting socket. The bridging condenser and the stopping condenser (C' and C_2 respectively of Figure 205) are controlled by the switches below the bulb. The three top-row knobs control an antenna loading coil, a secondary loading coil, and a coupling between the primary (antenna) circuit and the secondary circuit. The two lower knobs control an antenna tuning condenser and the secondary circuit tuning condenser. It will be seen that a special switch is used in connection with the loading coils so as to avoid dead ends when using only a portion of each coil.

The general appearance of the de Forest audion and three stage amplifier is shown in Figure 207. The lowest bulb is the detector, the remainder are audio frequency amplifiers. Each has its own filament current rheostat and its own dry cell plate battery variable in steps of 3 volts. The telephone can be plugged in at any stage as desired.

In Figure 208 is represented a general type of circuit devised by Dr. Meissner of the Telefunken Company. It differs from the preceding in the method of obtaining plate circuit outputs. Instead of inserting the telephone receivers into the plate circuit, the large inductance L_2 is

FIGURE 207—de Forest audion and three
stage amplifier.

placed in this circuit, and the alternating potential differences appearing at its terminals cause currents to flow in the tuned circuit $L_3 C_5$ which is coupled to L_2 by the condensers C_3 and C_4. The right hand bulb serves as a detector and amplifier, and finally delivers audio frequency currents to the telephone T. Another form of receiver of the Telefunken Company, devised by Count von Arco and Dr. Meissner is represented in Figure 209 It will be seen that this differs from Mr. Armstrong's circuit of Figure

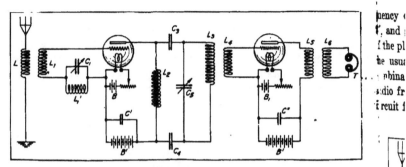

FIGURE 208—Telefunken Company-Meissner receiving system.

203 only in the mode in which the plate circuit output is delivered to the receivers.

The receiving set used with the Marconi Company's radiophone transmitter shown in Figure 153, page 156, and described in conjunction therewith is indicated in Figure 210. The grid circuit is coupled directly to the antenna circuit through a portion of the antenna inductance L and the inductance L'. The grid circuit is also coupled regenerative to the plate circuit through the coupling between L' and L''. The plate circuit is tuned by means of the condenser C'. The plate battery B has a voltage of 200 and the resistances R_1 and R_2, which limit the plate current, are each 2,000 ohms. The battery and its associated resistance are shunted by the condenser C'' which passes the amplified radio fre-

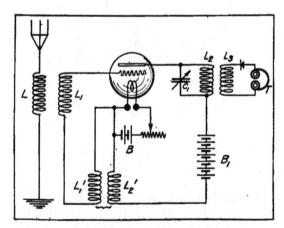

FIGURE 209—Telefunken Company-Arco-and-Meissner
receiving system.

uency current. The filament of the bulb is lit by the 6-volt battery
´, and grounded through the potentiometer resistance R'. The output
f the plate circuit is drawn from the condenser C' across which is placed
ιe usual crystal detector, auxiliary potential, and telephone receiver
ιbination. In other words, the system shown consists of a regenerative
:dio frequency amplifier combined with an ordinary crystal rectifying
rcuit for utilising the amplifier output.

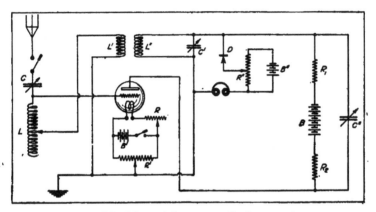

FIGURE 210—Marconi Company radiophone receiver.

Passing to the work of the Western Electric Company, we consider
irst some of the tubes developed by the engineers of that company and
heir method of construction. One type of tube, due to Mr. H. J. van
ι.r Bijl, is shown in Figure 211. Herein the objects are to keep the
ιlanes of the grid and filament close together (for high amplification)
ιnd to avoid undue tensions on the filament. As will be seen from the
ωwer portion of the figure, the filament is threaded to and fro on
he flat mica support, passing alternately from one side of the mica to
ιe other. The grid and its supporting frame are mounted close to the
ιιca, and are preferably arranged as shown in the lower portion of the
ιgure, that is, with the grid wires not crossing at the portions of the
ˁιa where the small vertical portions of the filament are exposed on
ωat side. In another form of tube, due to Mr. A. McL. Nicolson, and
˙epresented in Figure 212, it is attempted to secure "efficient control"
ιy twining the grid wire around the filament, separating them only by a
ιon-conducting or dielectric film. Nickelous oxid is recommended for the
ιurpose. In the form shown in Figure 212, the grid wires 1 are coated
ιith nickelous oxid, and around them are twined the filament wires 2.
Γhe plates 3 are situated as usual.

A type of Western Electric Company amplifier or "repeater" tube is shown in Figure 213. This type is due to Messrs. A. McL. Nicolson and E. C. Hull. The distinctive feature thereof is the twisted platinum filament 2, which is coated with metallic oxids. It is made by dipping a platinum ribbon having a width of say 0.3 mm. (0.012 inch) and a thickness of 0.05 mm. (0.002 inch) in chromic or nitric acid, washing it in water, and then in a strong solution of ammonia. After this thorough cleansing, it is heated to incandescence to see if it has any defects. It may then be dipped in a trough filled with dilute strontium hydroxid and thereafter dried at 100° C. (212° F.) by a current of 1.4 amperes. After

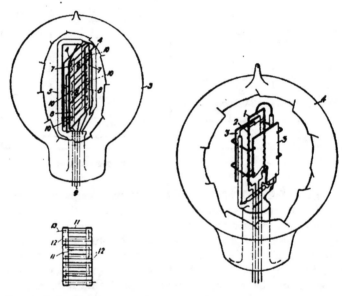

FIGURE 211—Western Electric Company-van der Bijl amplifier tube, 1915.

FIGURE 212—Western Electric Company-Nicolson amplifier tube, 1914.

four such coatings, the filament is heated to incandescence for a few seconds to harden the oxid film. It is next coated with barium resinate melted at a temperature of 600° C. (1,100° F.), and given four coats thereof as before except that it is heated for a few seconds to incandescence after each coat or two. The entire process thus far mentioned is then repeated, thus giving four sets of four coats of the oxid or resinate in all. The filament is then kept at incandescence for about 2 hours at 800° C. (1,470° F.) to ignite the resinate. The resulting film of strontium and barium oxids on the filament is smooth and tough and

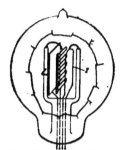

FIGURE   213 — Western
E l e c t r i c Company,
Nicolson-and-Hull
amplifier tube,
1914.

gives high electron emission at comparatively low temperatures, thus
tending to give a long filament life
in use. Tubes of this sort, but with
a grid and plate at each side of
the filament, are widely used by the
Company.

One type of receiver used by
the Western Electric Company con-

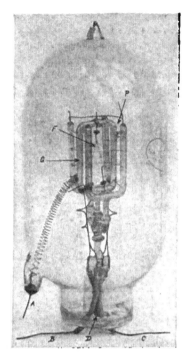

FIGURE   214 — General   Electric
Company-White small pliotron
amplifier.

sist of a number of steps. The antenna circuit is coupled to a radio
frequency regenerative amplifier, much like that shown in Figure 202 in
most respects except that the output is obtained by coupling in the plate
circuit to a fairly large impedance as in Figure 208. The next circuit
is a detector circuit, also provided with regenerative coupling. The output of this step passes into a two-step audio frequency amplifier with
inductive coupling between the steps. The final output is inductively
coupled to a balanced receiver for reducing the relative intensity of
strays, and devised on somewhat the same general lines as that shown
in Figure 215 except that three-electrode tubes are used.

The General Electric Company has constructed a number of different types of pliotron amplifier tubes, one type of which intended for
relatively small outputs is shown in Figure 214. The grid G is made
of very fine wire wound on a glass frame. Inside the grid is the "V"
or "W" shaped filament F. The plate P is not a solid plate of metal,
but consists of a zig-zag wire supported on wire supports placed appro-

priately in two "U" shaped glass frames. The filament leads are *B* and *C*, the plate terminal *D*, and the grid terminal *A*. The whole structure is carefully built and exhausted to an extremely high vacuum at which "pure electron" effects are obtained. Under these conditions it has been found possible to obtain extremely high audio and radio frequency amplifications without using regenerative circuits in connection with

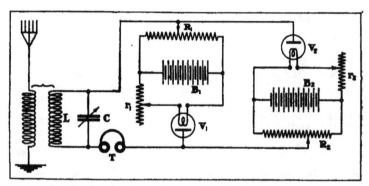

FIGURE 215—Marconi Company-Round balanced valve receiver.

the bulb, that is, without any other coupling between grid and plate circuits than the small capacitive coupling which necessarily exists within the bulb.

Mr. Alexanderson has shown several methods whereby a number of such tubes may be used in cascade, each giving radio frequency amplification. It is claimed that the selectivity of the resulting system is high, rising in geometric proportion to the number of steps.

(g) STRAYS.

We have previously considered to some extent "Stray Interference in Radio Telephony," page 18.

It remains to consider some of the technical expedients at present available for reducing somewhat the disturbing influence of strays. The chief of these are:

(1) Loose coupling between the antenna and receiver.

(2) Sharp tuning with circuits of fairly low damping.

(3) Beat reception (which is sometimes not readily applicable to radio telephony).

(4) Balanced crystal or valve detectors, which prevent excessive crashes of sound from reaching the ear.

(5) Special methods, given below.

The first three of these methods are commonly known. A simple circuit diagram illustrating the balanced valve receiver (as due to Mr. H. J. Round of the Marconi Company of England) is given in Figure 215. Here LC is the secondary of an ordinary receiver, and T a telephone receiver. The two hot-electrode rectifiers (Fleming valves) V_1 and V_2 are connected as shown. The batteries B_1 and B_2 serve not only to light the filaments through the appropriate controlling resistances r_1 and r_2 but also to provide a supplementary potential difference in the plate circuits through the potentiometer control resistances R_1 and R_2. It is well known that the curve connecting excitation (that is, incoming signal current) and response (that is, rectified current) depends, in

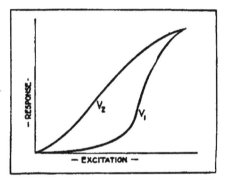

FIGURE 216—Valve characteristics with
different auxiliary potentials.

these valves, on the supplementary potential in the plate circuit. Hence we arrange that one of these valves shall have a favorable value of this potential, giving it high sensitiveness for weak signals. This will be valve V_2, and its sensitiveness curve is shown in Figure 216. The other valve, V_1, is run with a low supplementary potential, so that its sensitiveness for weak signals is very low. For extremely loud signals, however (because of the current saturation effect) its response is no less than that of V_2. It will be noticed that the valves V_1 and V_2 are connected in opposition or differentially so far as the receiver T is concerned. Hence weak signals will be readable since V_1 will not neutralise V_2 in this case. Strong crashes due to strays will affect both valves equally, and hence will not be heard in T.

An alternative scheme, proposed by Dr. L. W. Austin, is to connect a silicon-arsenic detector direct between antenna and ground as a shunt to the primary of the inductive receiving coupler. This detector is claimed not to affect weak signals, but to become conductive for ex-

tremely powerful disturbances, thus shunting them to ground and pro-
tecting the ear sensitiveness of the receiving operator.

A Dieckmann electrostatic shield for a flat top antenna is shown
in Figure 217. The purpose of such a shield is to prevent the electrostatic field of the earth or of the atmosphere from reaching the antenna,
by an action similar to that of a Faraday cage. At the same time, the
shield must be so arranged that the incoming electromagnetic waves
can pass through readily, as in the case of a Hertzian polarizing parallel-

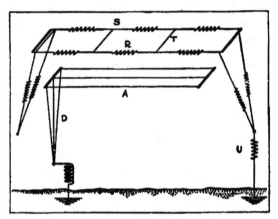

FIGURE 217—Dieckmann shielding cage for stray reduction.

wire screen. In Figure 217, A is the actual antenna with its down lead
D. The actual shielding wires are T and those parallel to it. The wires
R and those parallel to it are merely equalising connections, and include
inserted resistances so that the entire shielding system is aperiodic; that
is, incapable of being set into resonant oscillation by the incoming
energy. This is an obvious necessity. The shielding system is grounded
at U through a large inductance or resistance. In practice, Dieckmann
found that the reception was *louder* when the antenna A was shielded
than when it was not (because of increased capacity when the shield
was present). Naturally, this type of shield protects strictly against
"static" but not against all strays, since distant electromagnetic waves
from terrestrial sources can pass through it. In practice, however, Dieckmann found it to be of marked assistance very frequently, a fact since
verified by other careful investigators.

The Author has suggested in the past the use of a completely
covered antenna wire, the insulator being smooth and non-hygroscopic,
thus preventing charged air and water particles from giving their

charges by contact to the antenna, with the resulting disturbance of reception. Such a method should be of assistance at times, though it would naturally not be nearly so efficient a protection as a Dieckmann shield, since it would fail entirely to guard the antenna against aperiodic sudden changes in the earth's electrostatic field.

We shall now consider some further points in connection with strays and radio telephony. To begin with, it must be noted that with the modern sensitive receivers (e. g., the regenerative vacuum tube receivers), heavy strays do much more damage than to act merely as incidental noises. They break up the incoming sustained wave trains so far as the receiving system is concerned and thus prevent resonance phenomena in circuits of small (or even zero or negative) damping from being fully utilised. In addition, some types of detectors (e. g., sensitive crystals or gas-containing bulbs) may be "paralysed" by heavy strays and take some little time to regain their sensitiveness. Even very high vacuum tubes may show this effect, since a very powerful stray impulse may charge the grid negatively to such an extent that the plate current will be practically cut off until the grid charge escapes by the normal leakage. Then, too, the ear will be so shocked by heavy bursts of strays that it will take some little time to regain its normal sensitiveness.

The Author has found, in connection with some tests in radio telephony over fairly considerable distances, that really good reception requires that the signal audibility shall be 3 times that of the strays, and that the strays shall not occur continuously even then but intermittently, and not more frequently than say once or twice per second. Fair reception can still be accomplished even if the (intermittent) strays are 2 times as strong as the signal. Reception becomes difficult if the strays are 5 times as strong as the signal, and almost impossible if the ratio is greater, particularly if the strays are continuous.

Dr. C. J. de Groot has given in the "Proceedings of The Institute of Radio Engineers" for April, 1917, a classification of the strength of tropical strays (with a crystal detector and normal Telefunken receiver) and the signal strength necessary for telegraphic communication through such strays. The number of the class is on an arbitrary scale.

0. No disturbance. (Signal strength would be a few times audibility.)
1. Weak strays. Requiring a signal of 10 times audibility.
2. Medium strays. Requiring a signal of 20 to 30 times audibility.
3. Strong strays. Requiring a signal of 60 times audibility.
4. Heavy (or very heavy) strays. Requiring a signal of 250 to 500 times audibility.

5.   Overwhelming strays or thunderstorms. This case occurred only for an hour or two during the very worst days in the least favorable part of the year. Reception under such conditions is not possible.

Dr. de Groot showed that musical spark signals could be read through strays that had (at least intermittently) an audibility nearly 70 times as great as that of the signal and an audibility at almost all times of 10 or 20 times that of the signal. This he imputes to the remarkable selective sensitiveness of the ear to musical tones. This advantage is not present to the same extent in radio telephony.

Dr. de Groot further classified strays into three classes electrically and gave the details of the production, nature, and elimination of each class.

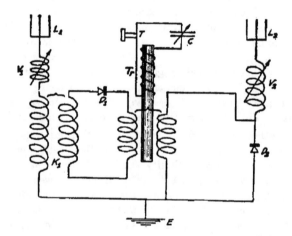

FIGURE 218—de Groot audio frequency compensation
method for the elimination of periodic strays.

Type 1. These are strays originating in nearby thunderstorms, and they have only a short range. They are found to be of periodic character (i. e., decadent wave trains of definite period and decrement). They are heard in the receiver as loud, widely separated clicks, and may be eliminated by audio or radio frequency compensation. The arrangement given by de Groot for this purpose is shown in Figure 218 and is explained by him as follows: "Two receiving antennas, L_1, L_2 of the same shape and dimensions were installed near enough together (10 or 20 meters or 30 to 60 feet apart) to make them respond in the same way to strays. (For the aperiodic disturbances this distance could be easily increased, but for periodic disturbances the distance of separation must be small compared to the wave length of the strays, in order to get the induced e.m.f.'s in phase). On the other hand the antennas must be placed sufficiently far apart so that the signals set up in the one which is made aperiodic (L_2) shall not cause currents in the tuned antenna (L_1). One of the antennas, L_1, is tuned to the incoming signal and coupled to the detector circuit D_1 in the ordinary way. The detector D_1 will rectify signals as well as strays and send the rectified current into the telephones; or, as in the case of the figure, into the differential transformer Tr. The antenna L_2 is tuned either to the same, or preferably to a longer wave-length, thus making it less sensitive to the signals and more sensitive to the long wave strays. The detector D_2 is switched directly into this antenna, thus making it aperiodic or nearly so. This arrangement makes it almost impossible to receive any distant signals on the antenna L_2, but loud signals on wave-lengths different from those to which L_1 is tuned and strays give a response that is nearly as loud as can be obtained on the tuned antenna L_1. The rectified current is sent to the same telephone mentioned before; or, as in the figure, to the differential transformer Tr. However, this second current from the aperiodic antenna L_2 is arranged to act in the opposite direction from that of D_1. The telephone T is either connected in series with D_1 and D_2; or, as in the figure, in a third winding of the differential transformer and in series with the condenser C to permit tuning to the spark frequency. Since D_2 does not respond to distant signals, there will be heard in the telephones the signals from D_1 only, whereas the strays rectified by D_1 and D_2 tend to compensate. By varying the coupling K_1, this compensation may be made complete.'' Dr. de Groot mentions that the detectors must both have the same characteristic and states that carborundum crystals with suitable auxiliary potential meet the requirements. The theory here given yielded very promising results when tested. For further details of this method, the reader is referred to the original article.

Type 2. These strays are associated with low-lying (electrically charged) rain clouds and are of very short range. Electrically, they are found to be intermittent uni-directional currents due to actual discharges to or from the antenna. They are audible as a constant hissing sound, and are eliminable by the Dieckmann electrostatic shield shown and explained in connection with Figure 217.

Type 3. These are most common or night strays and cause most of the interference with reception. They are believed to originate in the Heaviside layer or conducting portion of the upper atmosphere when this is subjected to the cosmic bombardment of small particles and comets. The range (with the receiver used) was several hundred miles,

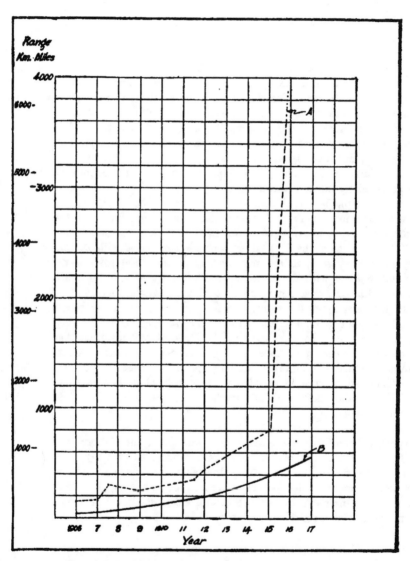

FIGURE 219—Yearly increase in range in radio telephony.

and these strays gave a continuous rattling noise. They were successfully eliminated by means of the Dieckmann cage of Figure 217.

Much valuable information on the daily and seasonal variation of strays is given in Dr. de Groot's paper. The reader is also referred to the Author's discussion on that paper for a further explanation of the Dieckmann cage.

The first approximation to the ratio of heavy summer strays to light winter strays is probably between 100-to-1 to 1,000-to-1 or even more.

(h) RANGE IN RADIO TELEPHONY.

As has been previously stated, the effective range of a radiophone transmitter depends on the loudness of strays at the receiving station; and consequently any method of reducing strays will increase the effective power of the transmitter in just that proportion.

In 1908, Mr. Fessenden, as the result of some rather elaborate analysis, reached the conclusion that the amount of power required to cover a given range radiophonically was from 5 to 15 times as great as that required to cover it radio telegraphically. It is certain, however, that some of the reasoning there given is not valid, and particularly that dealing with the greater amount of power required in radio telephony because of the relatively small amplitude of the higher harmonics in the human voice. It seems much more likely that about the same amount of power is required to cover a given distance by means of either system of communication excluding beat reception from consideration.

We have compiled from the material presented in this volume ,e maximum distance covered each year radiophonically. The data is given in the chart of Figure 219. It will be noticed that, practically speaking, radio telephony began in 1906 when a range of 160 miles (250 km.) was covered. It must be mentioned, however, that Fessenden had transmitted speech by a radio-frequent spark method a distance of 1 mile (1.6 km.) as early as 1900. The range increased fairly steadily at the rate of about 60 miles (100 km.) per year until 1915, when it took a sudden jump to the extreme range of 5,100 miles (8,000 km.). The dashed curve A shows this material clearly. We have, however, endeavored to distinguish between distance actually covered as an extreme achievement and the distance which could have been reliably covered with the apparatus available at any given time. The second curve B gives the range of probable reliable communication at any given year. It will be seen that this range has risen from about 40 miles (65 km.) in 1906 to about 500 miles (800 km.) in 1917. In fact, it is believed that with the equipment the performance oscillograms of which are given in Figure 200, page 202 (that is, the Alexanderson alternator-magnetic

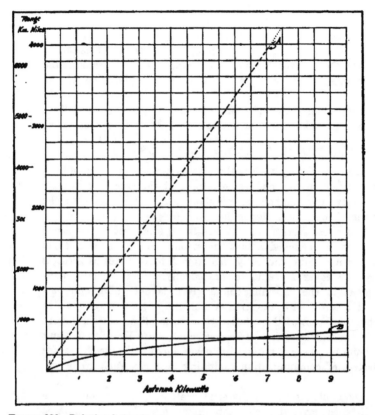

Figure 220—Relation between range and antenna power in radio telephony.

amplifier combination controlling 35 kilowatts) reliable overland telephony over 1,000 miles (1,600 km.) or more could be accomplished.
·We desire to emphasize particularly the distinction between "extreme
range" and "range of reliable communication." It is to be regretted
that we have had so much of the former type of achievement in radio
telephony and so very little of the latter. In view of the large ratio
between them, it is felt that only the latter type is of any real interest,
and that it only should be stressed hereafter.

In Figure 220 are given two curves connecting the range in radio
telephony with the *antenna kilowatts* (*not* transmitter input). These
curves are also based on the data given in this book. The upper curve *A*
gives "extreme ranges," and shows the following interesting facts.
With an antenna power of only about 0.5 kilowatt, 300 miles (500 km.)
can *sometimes*, though rarely, be covered. With 1 kilowatt, this rises to

600 miles (1,000 km.). At 10 kilowatts, it rises to 5,000 miles, (8,000 km.). The "range of reliable communication," given in Figure 220, curve *B* is very different. It will be seen that 1 antenna kilowatt will cover not over about 150 miles (250 km.) overland at the most desirable wave-length. For about 10 antenna kilowatts, this range rises to 500 miles (800 km.). The difference is very significant between curves *A* and *B*, and these curves cannot be brought closer together until the matter of stray elimination is settled. Even then, daylight and summer absorption of the electromagnetic waves will prevent the curves from being identical.

CHAPTER X.

10. RADIOPHONE TRAFFIC AND ITS REGULATION.
(a) DUPLEX OPERATION.

Any one who has compared normal telephone conversation with the irritating substitute provided by an ordinary speaking tube will realize the full necessity for duplex operation, i. e., simultaneous transmission and reception without the necessity for handling any switches or other devices when the speaker desires to listen, or vice versa. Experience teaches that sending-to-receiving switches lead to endless annoyance and confusion unless there is some skilled person standing next to the user of the radiophone to explain in detail how the switch is handled and to rectify errors of manipulation. While this latter procedure may be possible with a ship radiophone station, where the passenger desiring to telephone to land may put himself under the temporary guidance and instruction of the skilled radiophone operator, it would not be feasible on land since any system of land radio telephony must provide that calls may originate at any wire line subscriber's station, whether at his home or place of business. Since the land subscriber cannot, therefore, come to the radiophone station, there will be no opportunity to give him the necessary personal supervision and instruction.

A practical system of duplex radio communication (applicable to telephony) has been worked out by Mr. Guglielmo Marconi. The arrangement at the duplex sending and receiving station is shown in Figure 221. The transmitting antenna A is a long horizontal antenna, and directive (at any rate, for moderate distances and in reception). The main receiving antenna, A_1, is directive and parallel to the first. Both of these therefore point to the distant station. In addition to the main receiving antenna A_1, there is a balancing antenna A_2 so placed as to receive strongly from the transmitting antenna A but very little from the distant station. The distance CD in practice is anywhere from 25 to 50

230

miles (40 to 80 km.). A telegraph or, in our case, telephone line connects the stations. The receiver at *BD* is so arranged that it is coupled
to the coils in both antennas A_1 and A_2 *differentially*. By suitable
adjustment, it then becomes possible to cut out completely the signal
from *A* while retaining the signal from the distant transmitter almost
undiminished. Thus simultaneous transmission and reception become
possible.

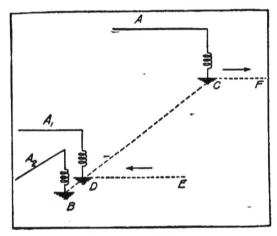

FIGURE 221—Marconi duplex station.

Another method, due to Mr. Fessenden, is indicated in principle
only in Figure 222. The four batteries *A*, *B*, *C*, and *D* are connected in
series assisting as indicated. Resistances R_1 and R_2 are inserted as shown.
Under these conditions, the points *X* and *Y* will be found to be at the
same potential, and a sensitive galvanometer connected across them will
show no deflection. Translated into the corresponding radio equivalent,
the actual arrangement is shown in Figure 223. The radio frequency
alternator *J* (in series with the microphone *M* or other controlling device)
sends current through the coils *H*, *G*, *E*, and *F*. There are thus induced
in the coils *C*, *D*, *A*, and *B* assisting currents. The resistance R_1 of Figure 222 is replaced by the artificial antenna R_1 of Figure 223, and the
R_2 of Figure 222 by the actual antenna of Figure 223. The four coils
in the antenna circuit correspond to the four batteries. As will be seen,
the points *X* and *Y* in Figure 223 are at the same potential so far as the
alternator is concerned, and a receiving set may be connected across
them when the arrangement is properly adjusted. This receiver will
respond (to some extent) to incoming signals. This entire arrangement,

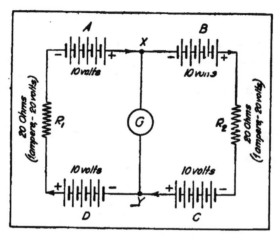

FIGURE 222—Principle of Fessenden duplex system.

while very ingenious, suffers from a number of practical disadvantages. To begin with, the exact balance is very difficult to secure and even more difficult to maintain because of changes in antenna and ground conditions. Furthermore, the points X and Y while at equal potential, are far above ground potential, and consequently capacity currents will flow from the receiving set to ground, disturbing the balance and giving

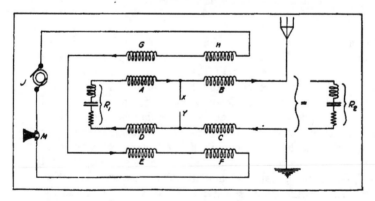

FIGURE 223—Fessenden duplex system for radio telephony.

false signals. In addition, at least half of the available energy will be lost in the artificial antenna R_1 and more than half of the incoming signal energy will be lost in all cases.

A somewhat similar arrangement, invented by Mr. J. H. Carson and assigned to the American Telephone and Telegraph Company, is shown in Figure 224. It will be seen that the secondary L' of the output transformer of the radio frequency alternator Q has two equal parallel load circuits. One of these is the path C_1, L_1, and the artificial antenna A_1, while the other is the exa. y similar path C_2, L_2, and the actual antenna A_2. The receiving set is coupled differentially to the two paths, and will

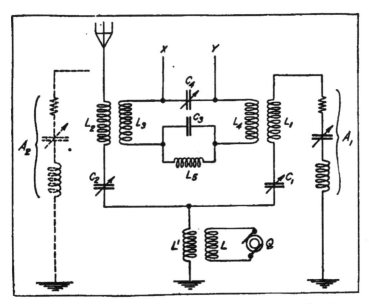

FIGURE 224—American Telephone and Telegraph Company-Carson system for duplex radio telephony.

therefore respond only to incoming signals. It contains the loop circuit L_5C_3 intended to cut down any unbalanced energy at the transmitting wave length which may chance to get into the receiver. This arrangement is subject to exactly the same defects as those pointed out in connection with Mr. Fessenden's above.

Another type of system intended to accomplish the same results as actual duplex working has been worked out by Dr. de Forest and along independent lines by some of the engineers of the General Electric Company. This consists of a voice-controlled relay which changes over the set from receiving to transmitting when speech is begun. A sluggish contact device (e. g., mercury in a capillary tube) is closed by the voice vibration or the exhaled breath and the set is then thrown, through the action of more robust relays, into the transmitting setting. The controlling device is usually located in or very near to the microphone transmitter. Other systems along similar lines have been proposed, all depending on changes caused by the voice or voice currents, but there is no data available for publication as the extent to which they are capable of practical application.

We will not discuss here such methods of duplex working as the commutator method, wherein the antenna is thrown in rapid succession

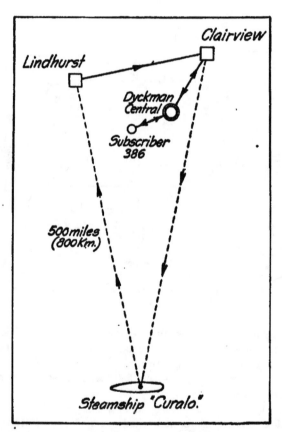

Figure 225—Typical ship-to-shore radiophone system.

from the transmitter to the receiver and back. While these may be suitable for telegraphy, they are obviously unsuited for telephony because of their almost certain destruction of the quality of the speech. Even if the commutation is done above audio frequency (which, in itself, is hardly very practical), the method would be open to grave objections.

(b) SHIP-TO-SHORE RADIO TELEPHONY.

The most casual consideration of the question of ship-to-shore radio telephony forces us to accept the conclusion that this vastly important system is dependent for its full development on the voluntary or enforced co-operation of the wire telephone companies. It is obvious that it is not possible to have a fairly large radiophone set at the home or office of every one who may at some time or other desire to speak with a person on board a ship, but that the land end of the conversation must be carried on from a large commercial radiophone station which automatically relays the speech out from the wire lines. Similarly the incoming speech from the ship must be received at the same or another radiophone station and there relayed back to the wire lines and thence to the subscriber. The procedure may be made clear from Figure 225. We will suppose that Mr. Frank Jones, whose wire telephone number in New York City is Dyckman 386, desires to radiophone to Mr. William Smith on board the steamship ''Curalo,'' some 500 miles (800 km.) at sea. Let us suppose that a duplex radiophone system (using, for example, the twin station Marconi plan given in Figure 221) is installed at the two towns near New York which have the assumed names of Clairview and Lindhurst. We shall take the transmitting station to be Clairview. At Clairview there will be a usual telephone connection, itself connected to a private telephone line between Clairview and Lindhurst. Lindhurst is the receiving station (with its balancing antenna as indicated in Figure 221). At Clairview, the incoming telephone line has inserted in it or across it the input side of a line amplifier which increases the energy of the speech current to the point of enabling control of the radiophone transmitter at Clairview. The wire line terminating in Lindhurst is there connected to the output side of an amplifier which increases the intensity of the incoming radiophone signals to the point where they can be sent through Clairview to the calling or called subscriber's station. On board the ''Curalo'' we have a moderately skilled operator, who among his other duties, listens for distress calls. Either the operator or one of the ship's engineers keeps the ship's radiophone set in order. We shall assume that the set on board the ship is not equipped for duplex work, though it is probable that eventually even the ship sets will be duplex. The change-over will be assumed to be accomplished by pressing down a push button when talking and releasing it when receiving, the push button circuit actuating some form of relay control switch which transfer from sending to receiving or vice versa.

We shall now proceed to give in detail the conversation between practically all the parties involved in the above call between Messrs. Jones and Smith. It is understood that this will be somewhat imaginative and subject to revision in details, though it is probably a fairly faithful impression of the actual procedure:

MR. JONES (on his wire telephone): Radio long distance, please.

OPERATOR (AT DYCKMAN CENTRAL): One minute, please. (She connects his line to the Clairview Radio Station line. The internal procedure at the central or centrals is here omitted.)

OPERATOR (AT CLAIRVIEW): Radio long distance speaking.

MR. JONES: I wish to speak to Mr. William Smith on board the steamship "Curalo."

OPERATOR (AT CLAIRVIEW): Mr. William Smith on board the "Curalo." What is your number?

MR. JONES: Dyckman 386, Mr. Frank Jones, the subscriber, speaking.

OPERATOR (AT CLAIRVIEW): Thank you. Hang your receiver on the hook. I will call you as soon as your connection is ready.

OPERATOR (AT CLAIRVIEW, talking out on the radiophone): Hello, Curalo. Hello, Curalo. Hello, Curalo. New York calling.

OPERATOR (ON "CURALO"): Hello, New York. "Curalo" talking.

OPERATOR (AT CLAIRVIEW): I want Mr. William Smith. Mr. Frank Jones of New York calling.

OPERATOR (ON "CURALO"): Mr. Frank Jones calling Mr. William Smith?

OPERATOR (AT CLAIRVIEW): Yes, please.

OPERATOR (ON "CURALO"): Hold the air, please. I will call Mr. Smith. (The operator on the "Curalo" then calls Mr. Smith to the radio cabin, and explains to him how to change from talking to listening by releasing the controlling push button. The method being learned, he resumes as follows):

OPERATOR (ON "CURALO"): Hello, New York. Mr. Smith is ready for you now.

OPERATOR (AT CLAIRVIEW, on wire line): Hello, Dyckman. Clairview calling. Give me 386 again, please.

OPERATOR (AT DYCKMAN): 386?

OPERATOR (AT CLAIRVIEW): Yes, please.

MR. JONES (at his wire telephone): Hello. This is Dyckman, 386. Mr. Frank Jones speaking.

OPERATOR (AT CLAIRVIEW): Mr. Smith is ready for you now. Go ahead please. (The operator at Clairview here closes the necessary amplifier circuits and takes a supervisory role only.)

MR. JONES: Hello, Mr. Smith. Jones calling.

Mr. Smith: Hello, Jones. This is certainly a pleasant surprise. How are . . . . .

It will be seen from a careful reading of the above that the procedure is no more elaborate than for any ordinary "particular person" long distance call. Furthermore, so far as the calling and called persons are concerned, there is no more difficulty or confusion than in any ordinary call. To verify this, the reader is urged to re-read Mr. Jones' and Mr. Smith's remarks above.

It need hardly be said that the system of charging for a radiophone conversation would be on the basis of time and not on the basis of words as in telegraphy. As to the extent of the charge, this might depend on several factors. To begin with, a somewhat deferred service corresponding roughly to "day letters" or even to "night letters" in ordinary telegraphy seems feasible at a considerably lower rate per minute. The season of the year and the distance over which the call has been made might also be factors of the situation, though to what extent only practical experience and the development of the art can determine.

There is one direction in which radio legislation properly conceived can greatly assist the radiophone field. This is by providing a system whereby every ship and its corresponding shore station have available not one or two, but a considerable number of wave lengths. These wave lengths, which should be designated by letters or numbers for the sake of brevity, would all be available for communication except those that were in actual use *near the receiving station*. That is, the receiving station, after listening for a moment, would dictate to the transmitting station the suitable wave length for communication without interference. Naturally all calling would be done on a common wave length which might be, for example, the present-day 600 meter wave. This system of a multiplicity of legal wave lengths and the choice of one of them for communication in accordance with traffic conditions at the receiving station has great possibilities, and should be carefully considered for future action by the International Radio Convention and the Governments of the world.

One further interesting possibility of radio telephony on board ship may be mentioned. A simple phonograph recording and reproducing device run by a small motor might be provided so that, in case the passengers and crews are forced to desert the ship after a serious accident, the phonograph can continue to repeat into the radiophone transmitter the necessary call for help, the name of the ship, its location, the type of accident, and the action taken by the passengers and crew. This would, to some degree at least, relieve the operator from the heroic, but frequently fatal, stand which up to the present he has always taken. With this simple device installed, he has at least the same chance of rescue as the other officers of the ship.

(c) LONG DISTANCE RADIO TELEPHONY.

This also must be accomplished with the co-operation of the wire telephone companies, and it is to be hoped that they will adopt a broad policy of co-operation with radio telephony in this regard. Since a large portion of the long distance radio telephony will be trans-oceanic (in which case wire telephony cannot come into competition), such an attitude on the part of the wire telephone companies will involve no inordinate sacrifice, and will, indeed, probably add very largely to the long distance land wire line tolls.

We shall give here also the sample procedure of a long distance radiophone call over the 5,500 miles (9,000 km.) between New York City and Buenos Aires. We shall suppose that Mr. Frank Jones of Dyckman 386 is calling Mr. J. Designate of Ciudad 762 in Buenos Aires. We shall assume now that Clairview and Lindhurst have the same functions as in the case described previously except that they are naturally provided with a much more powerful transmitter and a suitable receiving set. At Buenos Aires, the transmitting station is at the (assumed) town of Sol del Plata, and the receiver at the (assumed) town of Parina. The wire line connections, line amplifiers, and auxiliary apparatus are like those at Clairview and Lindhurst. Speech from Mr. Jones to Mr. Designante travels over the following route:

Mr. Jones at Dyckman 386—Dyckman central—(possible intermediate centrals not here considered)—Clairview—(by radio) Parina—(by wire) Sol del Plata—(possible intermediate centrals not here considered) —Ciudad central—Mr. Designante at Ciudad 762.

Speech from Mr. Designante to Mr. Jones travels as follows:

Mr. Designante at Ciudad 762—Ciudad central—(possible intermediate centrals not here considered)—Sol del Plata—(by radio) Lindhurst—(by wire) Clairview (possible intermediate centrals not here considered)—Dyckman central—Mr. Jones at Dyckman 386.

These paths are shown clearly in Figure 226. The detailed dialogue between all parties involved is here given. In addition, before each remark we give the elapsed time *in minutes and seconds* very roughly estimated:

0.00—MR. JONES (on his wire telephone): Radio long distance, please.

0.05—OPERATOR (AT DYCKMAN CENTRAL): One minute, please. (She connects his line to the Clairview Radio Station line. The internal procedure at the central or centrals is here omitted.)

0.25—OPERATOR (AT CLAIRVIEW): Radio long distance speaking.

0.30—MR. JONES: I wish to speak to Buenos Aires. A particular person call for Mr. J. Desigante, D-e-s-i-g-a-n-t-e, whose number is Ciudad 762.

0.45—OPERATOR (AT CLAIRVIEW): Buenos Aires, Mr. J. Desigante, D-e-s-i-g-a-n-t-e, of Ciudad 762. What is your number?

1.00—MR. JONES: Dyckman 386, Mr. Frank Jones, the subscriber, speaking.

1.05—OPERATOR (AT CLAIRVIEW): Thank you. Hang your receiver on the hook. I will call you as soon as your connection is ready.

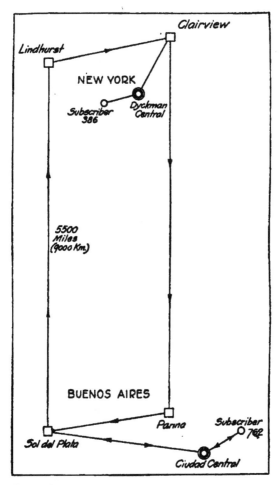

FIGURE 226—Typical duplex long distance radiophone system.

1.20—OPERATOR (AT CLAIRVIEW, talking out on the radiophone):
Hello, Buenos Aires. Hello, Buenos Aires. Hello, Buenos Aires. New
York, calling.

1.45—OPERATOR (AT SOL DEL PLATA, receiving through Parina):
Hello, New York. Buenos Aires speaking.

1.50—OPERATOR (AT CLAIRVIEW): I want Ciudad 762, Mr.
J. Desigante, D-e-s-i-g-a-n-t-e. Mr. Frank Jones calling.

2.15—OPERATOR (AT SOL DEL PLATA): Ciudad 762, Mr. J. Desigante, D-e-s-i-g-a-n-t-e. Mr. Frank Jones calling. Hold the air, please.
(Speaking on the wire line.) Hello, Ciudad. Sol del Plata calling.

2.35—OPERATOR (AT CIUDAD) (The internal procedure at the central
or centrals is here omitted): Hello, Sol del Plata. Ciudad speaking.

2.40—OPERATOR (AT SOL DEL PLATA): Ciudad 762, please.

2.42—OPERATOR (AT CIUDAD): 762?

2.45—OPERATOR (AT SOL DEL PLATA): Yes, please.

2.55—MR. DESIGANTE (on his wire telephone): Hello, this is Ciudad
762. Mr. J. Desigante speaking.

3.00—OPERATOR (AT SOL DEL PLATA): Mr. Frank Jones of New
York wishes to speak to you. Hold the wire, please. (By radiophone):
Hello, Clairview. Ciudad 762 is ready for you now.

3.20—OPERATOR (AT CLAIRVIEW): Thank you. Hold the air,
please. (By wire telephone): Hello, Dyckman. Clairview calling.
Give me 386 again, please.

3.30—OPERATOR (AT DYCKMAN): 386?

3.33—OPERATOR (AT CLAIRVIEW): Yes, please.

3.50—MR. JONES (on his wire telephone): Hello, this is Dyckman
386. Mr. Frank Jones speaking.

3.53—OPERATOR (AT CLAIRVIEW): Mr. Desigante is ready now. Go
ahead, please. (The operator at Clairview here closes the necessary
amplifier circuits and takes a supervisory role only.)

3.56—MR. JONES: Hello, Mr. Desigante. Jones speaking.

4.00—MR. DESIGANTE: Hello, Mr. Jones. Desigante speaking.

4.01—MR. JONES: About that shipment number 1167 of April 18th
on the "Curalo," I wanted to ask whether . . . .

As stated before, the charges on such telephone service might well take account of the time of day and of the season of the year.

There will be an interesting competition between very high speed telegraphy (possibly with automatic recording *and transcribing* apparatus) and radio telephony in connection with the normal transaction of business. It is too early to venture any predictions regarding the results of such competition. However, for personal communications there can be no doubt as to which form of communication will be preferred.

(d) FUTURE DEVELOPMENT OF RADIO TELEPHONY.

Now that the need for radio telephony is well recognized, we may confidently expect a very rapid development. Once a public demand is created, the technical advances required to satisfy that need must shortly follow.

Some interesting possibilities as to universal communication may be considered. So far as portable transmitters are concerned, it is unlikely. for some time to come that a man will be able to carry a radiophone set capable of communicating more than a few miles. Some new motive force, the apparatus for producing which has a very small weight per kilowatt delivered, must first be discovered. So far as reception is concerned, however, very sensitive and light receiving apparatus, capable of receiving messages from hundreds (or even thousands) of miles is imaginable. So that it should become ultimately possible to keep in immediate touch with the traveling individual regardless of his motion or temporary location. A great field of usefulness is thus opened to development.

The linking of the wire telephone and radiophone systems of a country will go far toward making it possible for travelers to keep in touch with their homes and business at all times, and for the people of one nation to know the people of far distant nations in a close and intimate fashion. By the use of a deferred or night radiophone service (analogous to the day letter or night letter of the wire telegraph companies), reasonably inexpensive communication of this type should become feasible since such service might be rendered at times of light load and would tend to maintain the steady usefulness of the station. As is well known, stations are most efficiently operated when the load is nearly always full and constant. Plant efficiency requires, therefore, that some sort of premium be put on utilization of the plant facilities at times of normally light load.

In conclusion, it may be stated that it is certain that radio telephony, properly fostered by the Governments of the world, must become ever more useful to humanity. From ship and shore stations, from aeroplane

and ground, from trains and depots, over forests and deserts, across
oceans and continents will pass the spoken word of man. We may justly
paraphrase President Elliot's splendid eulogy of another type of communication. His words apply with multiplied force to the radiophone of
the future. We may rightly term this instrument for speeding the voice
of man across space as :—

CARRIER OF NEWS AND KNOWLEDGE.
INSTRUMENT OF TRADE AND INDUSTRY.
PROTECTOR OF LIFE AT SEA.

MESSENGER OF SYMPATHY AND LOVE.
SERVANT OF PARTED FRIENDS.
CONSOLER OF THE LONELY.

BOND OF THE SCATTERED FAMILY.
ENLARGER OF THE COMMON LIFE.

PROMOTER OF MUTUAL ACQUAINTANCE.
OF PEACE AND GOOD WILL AMONG MEN AND NATIONS.

INDEX

All references to individuals, companies, equipment, circuits, or radio stations are fully listed in this index. With the exceptions of the names of companies, all topics will be listed in general under the NOUN to which reference is made. The numbers correspond to pages in the text. The following abbreviations are used: r.f.—radio frequency; a.f.—audio frequency; r.t.—radio telephony; w.t.—wire telephony.

243